T0234725

SpringerBriefs in Applied Sciences and Technology

SpringerBriefs present concise summaries of cutting-edge research and practical applications across a wide spectrum of fields. Featuring compact volumes of 50–125 pages, the series covers a range of content from professional to academic.

Typical publications can be:

- A timely report of state-of-the art methods
- An introduction to or a manual for the application of mathematical or computer techniques
- A bridge between new research results, as published in journal articles
- A snapshot of a hot or emerging topic
- An in-depth case study
- A presentation of core concepts that students must understand in order to make independent contributions

SpringerBriefs are characterized by fast, global electronic dissemination, standard publishing contracts, standardized manuscript preparation and formatting guidelines, and expedited production schedules.

On the one hand, **SpringerBriefs in Applied Sciences and Technology** are devoted to the publication of fundamentals and applications within the different classical engineering disciplines as well as in interdisciplinary fields that recently emerged between these areas. On the other hand, as the boundary separating fundamental research and applied technology is more and more dissolving, this series is particularly open to trans-disciplinary topics between fundamental science and engineering.

Indexed by EI-Compendex, SCOPUS and Springerlink.

More information about this series at http://www.springer.com/series/8884

Hamid Reza Rezaie ·
Mohammad Hossein Esnaashary ·
Masoud Karfarma · Andreas Öchsner

Bone Cement

From Simple Cement Concepts to Complex
Biomimetic Design

 Springer

Hamid Reza Rezaie
Ceramic and Biomaterial Division
Department of Engineering Materials
Iran University of Science and Technology
Tehran, Iran

Masoud Karfarma
Ceramic and Biomaterial Division
Department of Engineering Materials
Iran University of Science and Technology
Tehran, Iran

Mohammad Hossein Esnaashary
Ceramic and Biomaterial Division
Department of Engineering Materials
Iran University of Science and Technology
Tehran, Iran

Andreas Öchsner
Faculty of Mechanical Engineering
Esslingen University of Applied Sciences
Esslingen am Neckar, Baden-Württemberg
Germany

ISSN 2191-530X   ISSN 2191-5318   (electronic)
SpringerBriefs in Applied Sciences and Technology
ISBN 978-3-030-39715-9   ISBN 978-3-030-39716-6   (eBook)
https://doi.org/10.1007/978-3-030-39716-6

This Springer imprint is published by the registered company Springer Nature Switzerland AG
The registered company address is: Gewerbestrasse 11, 6330 Cham, Switzerland

Preface

Cement is a compound that can be found everywhere, in buildings, roads and also in bone surgeries. Its structure comes from simple interaction between two acidic and base compounds. Various compositions of polymeric and ceramic cements have been used as bone cement such as poly(methyl methacrylate), calcium phosphate, magnesium phosphate and calcium silicate. However, scientists want more than this simplicity and aim to improve its resemblance to natural bone by designing complex biomimetic structures. Their designs obey three main elements of the tissue engineering pyramid, including biomaterials, cells and bioactive molecules. In this book, the main concept of the cementation process and effective parameters on the process are introduced in Chap. 1. In Chap. 2, composition and structure of candidate biomaterials for this application are considered. In Chaps. 3 and 4, the authors show the attempts of different research groups to improve the ability of naked biomaterials in enhancing bone healing by adding cells and bioactive agents.

Tehran, Iran   Hamid Reza Rezaie
Tehran, Iran   Mohammad Hossein Esnaashary
Tehran, Iran   Masoud Karfarma
Esslingen am Neckar, Germany   Andreas Öchsner

Contents

Chapter 1
Cement Concept

1.1 Introduction

Cement is a substance that by hardening from plastic condition binds separate solid particles together. The substance was invented in ancient days, based on various evidence found from Roman and Persia Empires. Since this time, its properties have been improved, and its applications have been broadened to encompass different fields, such as engineering and medical fields. In medical surgeries, cement has been applied as augmentation agent of orthopedic implants and filler of bone defects. In this chapter, the authors focus on bone cement, its applications, properties, and effective parameters. In the end, the combination of tissue engineering strategy with outstanding features of bone cement is assessed.

1.2 History

In the ancient world, as confirmed by prehistoric megalithic structures, the structures were constructed without using any adhesive materials, only by arranging heavy stones on each other. Ancient Egyptians built their structures by unburnt bricks covered by the loam, Nile mud. Even, the Great Pyramids were not constructed by a thing like that we know as modern cement, in substitution, burnt gypsum was used as mortar in those structures. Babylonians and Assyrians applied bitumen to binding burnt bricks. As the first one, it can be pointed to the Greeks that mixed slaked lime with sands like as modern fashion. Romans also applied the same route as the Greeks, maybe borrowed from them. In addition, they improved the strength of the mortar by adding volcanic tuff, and in the lack of the stone, they used powdered tiles or pottery instead. The usage of this mortar was spread in the whole of territories of the Roman Empire. For example, the Pantheon can be mentioned (Fig. 1.1), as a building remained from the Empire. After the collapse of the Empire, it seems that the art of producing burnt lime vanished and the structure built after the time was

created of poorly prepared mortar. The trend continued until the twelfth century that
the quality of the mortar was improved again. It took until 1850 that gradually the
historical kind of cement was substituted by modern Portland cement [1].

Since 1850, cement manufacturing has been evolved significantly. Some of
these occurrences are as follows: inventing rotary kiln, transferring from wet
process to dry/semi-dry process, adding computer-controlled systems, and facing
environmental challenges [3].

In ancient Iran, Persia, an organic cement named "Sarooj" was used. The cement
was produced in two types, warm and cold. In the warm type, clay-contained lime-
stones were hammered to transform them to dust and then mixed with dung, straw,
and water. The mixture was kneaded for 1–2 days and then spread on the surface with
a thickness of 5–10 cm. The spreading material was cut into rectangular cuboids with
a dimension of 20 × 30 cm, called Casts. The Casts were left for three days to dry.
The obtained product, named Kheshtak, was put on brushwood and burnt. Finally,
the burnt compound was ground and used in buildings. The cold Sarooj was a mixture of slaked lime and ash, as adhesive materials, clay and sand, as fillers, different

kinds of fibril materials including wool, hair, straw and bamboo dust, as preventing crack formation, and egg white, as a strengthening agent. The application of Sarooj was observed in different historic buildings in Iran such as an ancient ziggurat (1250 BC) called Chogha Zanbil, water reservoirs called "Ab-Anbar," and Safavid bridges, i.e., "Si-o-Se Pol" [4–6].

In addition to the great performance of cement to produce bridges, buildings, pathways, etc., its application in the human body as bone cement has been considered for more than a half of century. To introduce this application, in the upcoming section, the history of cement application in the medical area would be discussed.

1.3 Medical History of Cement

In the modern era, by developing technology and improving the health issues confronted human, human life expectancy has been enhanced. Scientists attempt hard to cure diseases and introduce new approaches to decrease pain. Among them, human hard tissue, bone, is a great challenge because it cannot repair itself independently. So, many approaches, such as autograft, allograft, and xenograft, were invented to favor bone growth. In these methods, a bone substitution originated from the patient, other human or other species was inserted in a bone defect to guide bone growth. Nevertheless, some problems, including the lack of sources and immunogenicity issues, encouraged scientists to seek alternative methods. Using artificial substitutions and bone cement were among the proposed methods [7].

In medical applications, for the first time, a cement composed of plaster and colophony was applied by Themistokles Gluck, 1870, to fix an ivory total knee prosthesis. After understanding the polymerization mechanism of polymethyl methacrylate by Degussa and Kulzer, 1943, John Charnley (1958), a British surgeon, used polymethyl methacrylate to fix a total hip arthroplasty. The fixation application of polymethyl methacrylate was approved by the U.S. Food and Drug Administration (FDA) in the 1970s [8]. Nowadays, beside polymethyl methacrylate, a new kind of cement has also been commercialized, named calcium phosphate cement. In the next chapter, more information about the cement composition will be discussed.

1.4 Cement Applications

The health of bone can be threatened by different sources, including externally applied stress and some diseases. For example, osteomyelitis is a disease that can originate from hematogenous spread of bacteria that infect part of a bone, or vascular or neurological insufficiency [9]. A para-articular intraosseous cyst is the other one that can appear as a side-effect of arthrodesis or a deflection in the performance of chondrogenic cells to get cancerous [10]. A vertebral compression fracture is a disease caused by osteoporosis [11]. The pain caused by the mentioned diseases can be

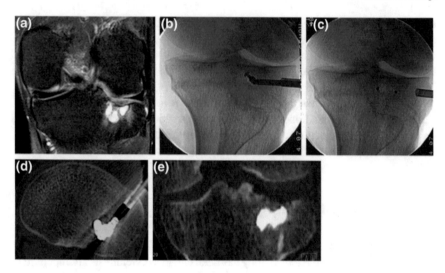

Fig. 1.2 **a** A cyst placed at the lateral tibial plateau, **b**, **c** applying a spacer and a balloon to create a cavity, and **d**, **e** injecting bone cement and evaluating the procedure (Re-printed from [10] by permission from Springer Nature)

Fig. 1.3 Short segment pedicle screw composed of four screws and two short rods (Reprinted from [13] by permission from AO Foundation)

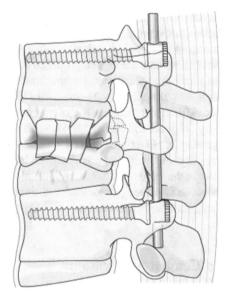

eliminated by applying bone cement. As can be seen in Fig. 1.2, the position of the infected bone was evaluated by CT scan and fluoroscopic methods. Then, the position was evacuated by a spacer, and the defect cavity made uniform via an inflated kyphoplasty balloon. Finally, the cavity was filled by bone cement and the procedure was controlled by the mentioned methods [10, 11]. In some cases, such as vertebral compression fracture, an additional device is required, specially when ceramic bone cement is used. The device is named short segment pedicle screw and shown in Fig. 1.3. In this method, the device preserves the brittle bone cement from applying external stresses, and the injected cement reduces the applying stresses on the device that can result in fatigue fracture [12, 13]. Bone cement can also be used in a non-loaded bone defect, for example, in the defect placed in the cranio-maxillofacial region [14]. Furthermore, bone cement is injected around arthroplasties to improve their integrity with surrounding bone tissue. In the case of cementless arthroplasties, the occurrence of wear creates debris that can cause chronic pain and the death of the surrounding tissue [15].

1.5 Why Bone Cement?

Nowadays using various arthroplasties, as a bone substitution, is common in the treatment of bone diseases. However, what feature does make applying bone cement an outstanding and preferable method? When a surgeon uses a conventional arthroplasty, first he/she should evaluate the position of a defect by imaging techniques. Then, based on determined dimensions, the arthroplasty should be sized, and after incision on the surrounding tissues, the device is placed in the defect site. The created wound prolongs the curing time and increases the risk of infection and feel of post-surgical pain. Nevertheless, when a surgeon applies an injectable cement, he/she uses methods that create a minimal incision, and without any predetermined knowledge, he/she can fill any defect placed even in a remote and inaccessible site. Hence, the curing period would be reduced, and the patient can resume his/her career in a short time [16, 17].

1.6 Bone Cement Properties

An injectable material to be known as an acceptable bone cement should possess some properties discussed as follows. The first one is the easiness of preparation. The preparation step is one of the most effective steps in the final properties of bone cement. Because a surgeon is confronted with many stresses during surgery and the skill level of each surgeon is different, the preparation should be so that any surgeon can easily mix precursor powders/liquids/pastes and create repeatable final properties. Usually, bone cement is produced by mixing a mixing a precursor powder and an aqueous liquid. However, other types of precursors that make preparation easier also are released in the market, including multiple liquid/pasty phase cement, pre-mixed cement, pre-mixed frozen slab. The first one is composed of a non-aqueous liquid or paste containing dispersed precursor powders and an aqueous liquid. In the second type, precursor powders are dispersed in a non-aqueous liquid that can be dissolved in water. The last one is a cement past whereas its setting process is prevented in a temperature lower than −80 °C [18].

The second property is the setting temperature and the pH. The setting temperature should be limited in the range that can be tolerable with body components. If the temperature goes beyond the range, thermal necrosis would appear in a tissue. Some studies showed that there is a relationship between the applied temperature and the time that the surrounding tissue can be preserved from necrosis. By increasing 1 °C, the duration of time that surrounding tissue can tolerate decreases with a factor of 2 [19–21]. Different cell types and tissues possess different thermal sensitivities, from 43 to 57 °C. For example, thermal necrosis in bone tissue and nerves can appear when they face with the temperature of 50 and 45 °C for the duration of 1 and 30 min, respectively [22, 23]. The pH is an essential item in the physiology of the human body and usually maintains in the range of 7.35–7.45. The range is crucial for the activity of enzyme systems, blood coagulation, muscle contraction, cell metabolism, etc. [24]. Although the pH is controlled by many different body buffer compounds, respiratory system, kidney function [25, 26], the pH of cement setting process should not be so to have an adverse effect on body functions.

Setting time and cohesion time are other important features of bone cement. As shown in Fig. 1.4, the setting process of bone cement can be divided into some periods. The first one ends in cohesion time. During this period, after mixing precursor powders/liquids/pastes, cement paste should obtain sufficient strength against blood and body fluid. The second period continues to the initial setting time that the period is the best time for injection of paste. The last period ends with the final setting time. At

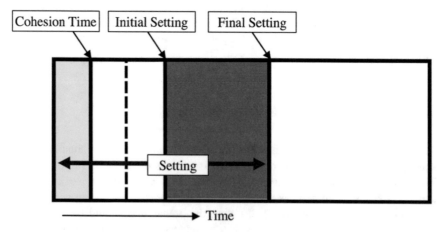

Fig. 1.4 The schematic of important periods in the setting process of bone cement (Adapted from [27] by permission from Sage)

the end of the period, the paste should achieve an acceptable initial strength that lets a surgeon accomplishes the surgery and closes the wound. The mentioned periods should be so that either a surgeon has enough time for preparation and injection of the paste, or the cement paste gains sufficient mechanical properties very soon. Usually, the initial setting time in dental and orthopedic applications is considered 3 and 8 min, respectively, and the final setting time in both of the applications is considered at most 15 min [27, 28].

Another considerable property of bone cement is its ability to be injected. For this feature, two items must be considered, including viscosity and filter pressing. In terms of viscosity, the paste should be shear thinning, i.e., when shear stress is applied on a highly viscous paste, the interparticle bonds are broken, the viscosity is significantly reduced, and the paste is easily injected from a syringe. Then, when the paste fills a defect, and the stress is eliminated, the structure of the paste rearranges, and the paste regains its high viscosity [29]. This feature from one side favors injection of the paste and from the other side maintains the paste in its position. On the other hand, filter pressing should also not be happening during injection, i.e., powder to liquid ratio and particle size distribution should not be significantly varied. Filter pressing appears when the required pressure for liquid filtering through interparticle space is lower than the required pressure for paste injection. The event deflects the final properties of paste from the pre-determined purpose [30].

When paste is set, its final mechanical properties must conform to the mechanical properties of natural bone tissue [31], as indicated in Table 1.1. Sufficient mechanical properties of injected cement allows rapid strengthening of a filled defect zone and early mobility of a patient [17]. If the injected cement cannot tolerate applied stresses, it would be destroyed and create some debris which impresses the health of surrounding tissues and may result in an infection [32, 33].

According to the second and third generations of biomaterials, they should be osteoconductive, osteoinductive, biodegradable and mimic the feature of goal tissue. So, the final cement is better to be porous to allow ingrowth of surrounding tissues. Biodegradability is another preferred feature that should conform to the growth rate of surrounding tissues to maintain interface integrity. Moreover, last but not least, cement should be osteoconductive and osteoinductive to conduct and induce bone to grow through the injected cement and substitute its position. Other features such as

Table 1.1 Properties of different human bone tissue (Reprinted from [31] by permission from Elsevier)

	Compressive strength (MPa)	Flexural strength (MPa)	Tensile strength (MPa)	Modulus (GPa)	Porosity (%)
Cortical bone	130–180	135–193	50–151	12–18	5–13
Cancellous bone	4–12	NA	1–5	0.1–0.5	30–90

NA indicates data not available

low cost and being opaque against radio waves, to favor detecting implant variation, are also important [17].

1.7 The Setting Process

Because of cement compositions include the ceramic base and the polymer base, two mechanisms of setting consisting of polymerization and cementation are considered. In the polymer base cement, polymerization and crosslinking are the dominant mechanisms. Polymerization occurs in three steps including initiation, propagation, and termination. In the first step, by applying heat, electromagnetic radiation, or through electrochemical approaches, the added small trace of an initiator is fragmented into free radicals. The produced free radicals react with unsaturated monomers to create new free radical species. In the next step, the reaction between the new free radical species and monomers increases the molecular size of the free radical. In the end, the propagation continues until the free radicals react together to produce a single bond, the reaction is called combination, or to produce either a saturated or an unsaturated products, the reaction is named disproportion [34].

Usually, ceramic cement is produced from an acid-base reaction. During the reaction, a base compound is dissolved by an acid solution and releases construction ions. By increasing the concentration of ions in the liquid solution, the ions accumulate in an arranged array and create a complex compound. By increasing the concentration of the complex compound, the liquid transforms to a gel structure and crystalline, semi-crystalline and amorphous compound precipitate from the gel [35]. Figure 1.5 shows the released heat during the cementation process. The diagram is divided into four zones including initial reaction, slow reaction, acceleration period, and deceleration period. At the first period, a massive amount of heat releases in a short time, and then in the next zone, the amount severely reduces. When precursor powders are

Fig. 1.5 The variation of released heat during cementation process (Reprinted from [36] by permission from Elsevier)

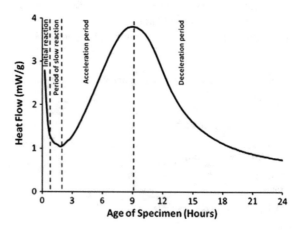

imposed to a liquid whereas its ionic concentration relative to the precursor powders is far from the equilibrium state, the high potential energy let the powders rapidly dissolve in the liquid, and the dissolving process releases a high amount of heat in a short time. Then by approaching to the equilibrium state, the reaction rate decreases. The period of slow reaction continues until the concentration of the liquid reaches to a supersaturation level, i.e., the required potential energy for nucleation and growth of new substances is obtained. Because of the consumption of the released ions while creating the new phases, the ionic concentration of the liquid distances from the equilibrium level and the dissolving reaction again accelerates to compensate the deficiency. In the last period, the released heat decelerates due to four reasons: (1) by the formation of the new compounds on the precursor powders and increasing of their thickness, the diffusion of ions that are effective on the dissolution of the precursor powders, and nucleation and growth of the new compounds is restricted; (2) all of the fine precursor powders are dissolved, and only the coarser ones remain; (3) the deficiency of space prevents the new compounds from natural growth; and (4) the shortage of remained water as one of the reactants [36, 37].

1.8 The Effective Parameters

To control different features of cement, three important parameters should be considered. In the following, these three parameters, including powder to liquid ratio, particle size/molecular size, and reactant reactivity, will be discussed.

1.8.1 Powder to Liquid Ratio

A decrease in the ratio of powder to liquid improves wetting of precursor powders and decreases their agglomeration. In addition, the presence of high amount of water prolongs the setting time of the paste cement and enhances its injectability [38]. Wu et al. [39] evaluated the variation of the powder to liquid ratio on calcium-magnesium phosphate cement. According to their results, an increase in the powder to liquid ratio from 4 to 8 g/ml decreased setting the time of the cement paste from 12 to 6 min, and also reduced its injectability. The injectability reduction was so that at the highest powder to liquid ratio, the paste cannot be injected. Habib et al. [40] to eliminate the setting time effect on injectability and understand the pure effect of the powder to liquid ratio on it, replaced α-tricalcium phosphate, as a soluble precursor powder, by a non-soluble one, β-tricalcium phosphate. As shown in Fig. 1.6, an increase in the powder to liquid ratio decreased the percentage of extruded the paste from a syringe and increased the required force for injection.

Because the presence of voids can severely affect the final mechanical properties of cement, determining an optimum ratio of powder to liquid, that preserves an equilibrium between sufficient reaction of precursor powders and void formation, is

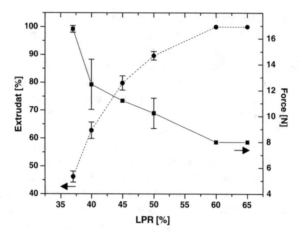

Fig. 1.6 The effect of liquid to powder ratio (LPR) on the percentage of extruded paste and the required force for injection (Reprinted from [40] by permission from Elsevier)

very important. For example, Moseke et al. [41] reported that an increase in powder to liquid ratio from 1 to 3.3 g/ml increased the density of the final cement from 1.58 to 1.9 mg/mm^2 and enhanced their compressive strength from 19.72 to 45.6 MPa. Based on research done by Espanol et al. [42], the formed porosity in cement can be divided into two classes, as shown in Fig. 1.7: (1) the voids placed between the formed needles or crystallites of the new compounds; and (2) the interparticle space remains between precursor powders when the new compounds coated the powders. By an increase in the powder to liquid ratio, the local supersaturation occurring in liquid solution rises the amount of nucleation of the new compound and decreases

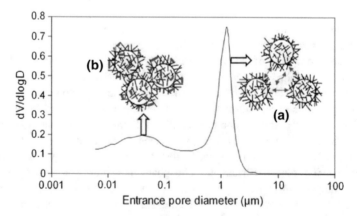

Fig. 1.7 Two class of voids in cement structure, **a** the voids between precursor powders covered by formed new compound, and **b** voids between formed crystallites (Reprinted from [42] by permission from Elsevier)

the dimension of formed crystallites on the precursor powders. Therefore, the size of the first kind of voids is decreased. On the other hand, the interparticle space filled with the liquid reduces and in consequence, the dimension of the second type of voids also decreased. So, the synergic events improve the mechanical properties of cement.

The ratio of polymer to monomer can be valid on the final mechanical properties of the cement. The higher ratio increases the viscosity of polymer/monomer mixture and causes incomplete polymerization which results in higher porosity and lower mechanical properties [43]. Increasing the amount of the crosslinking agents can cause producing a structure with longer crosslinks that reduce the density of the structure and its mechanical properties [44].

1.8.2 Particle Size/Molecular Size

An optimum size should be determined for precursor powders. If they are too coarse, the reactivity of the powders would reduce, and the coarse powders would block the passage of cement paste during injection. The high reactivity of the fine powders accelerates the setting process of cement and produces coarse agglomerated particles that can also decrease the injectability of cement paste. Figure 1.8 shows the effect of fine and coarse precursor powders, 1 and 10 um, respectively, on the injectability of the cement paste [45]. In this research, two types of precursor powders including α-tricalcium phosphate, a soluble powder, and β-tricalcium phosphate powder, an insoluble one, with the liquid to powder ratio of 0.35 and 0.45 ml/g were evaluated. The fine α-tricalcium phosphate powders with a low amount of water cannot be injectable. By comparing the mentioned powder by the fine β-tricalcium phosphate powders, it is understandable that the reactivity of the fine powders was the cause

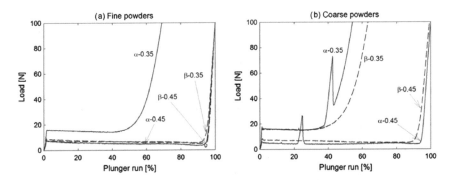

Fig. 1.8 Variation of required force for injection of the cement paste containing **a** fine (1 um) and **b** coarse (10 um) precursor powders. The powder compositions were α-tricalcium phosphate and β-tricalcium phosphate mixed with water by the liquid to powder ratio of 0.35 and 0.45 ml/g (Reprinted from [45] by permission from Elsevier)

Fig. 1.9 Variation of the mechanical properties of three kinds of cement versus particle size reduced by grinding method (Reprinted from [46] by permission from Elsevier)

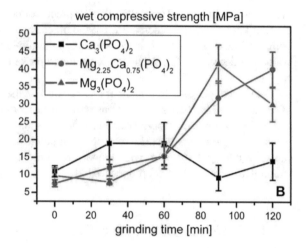

of preventing the paste from injection. The coarse powders in both of the types of powders blocked the passage of the syringe and affected the injectability of the paste.

The particle size of the precursor powder is also effective on the final mechanical properties of cement. Because of the high reactivity of fine powders, the cement set in a short time and its operator does not have enough time for preparation. On the other hand, coarse powders do not let that sufficient portion of the powders to contribute to the setting reaction and creates suitable local supersaturation of released ions. Therefore, both occurrences can decrease the final mechanical properties of cement. As shown in Fig. 1.9, Klammert et al. [46] reduced the size of the precursor powders by increasing the time of the grinding process. According to the composition of the used precursor powders that makes different reactivity among them, the powders only in a defined range can be used to obtain optimum compressive strength.

Abdul-Samad and Jaafar assessed the effect of polymer molecular weight on the setting of polymethyl methacrylate bone cement. The powder with lower molecular weight set in a shorter time because the penetration of monomers into the high molecular weight particles was difficult [43]. The research done by Peter [44] showed that higher molecular weight of polypropylene fumarate could increase the crosslinking temperature. The effect originated from higher viscosity of the polymer solution that increased the autoacceleration of polymerization. However, the higher molecular weight can define heat release during the crosslinking process. As they reported, by increasing molecular weight, the mobility of the polymer chains reduced and steric hinderance eliminated a part of double bonds to incorporate in the crosslinking process.

The gelling time of a polymer is also affected by the molecular size of a polymer. The higher molecular size needs lower crosslink to produce a structure, so the polymer gels in a shorter time [44].

1.8.3   Reactant Reactivity

The reactivity of precursor powders is affected by the particle size (mentioned in the previous section), the crystallinity degree of the powders, composition, and additive materials. In the case of injectability, Montufar et al. [45] compared amorphous and crystalline precursor powders of α-tricalcium phosphate and reported that lower reactivity of the crystalline one enabled injectability of the produced cement paste. Other researches [47, 48] showed that adding gelatin and soybean to α-tricalcium phosphate powders decreased the powder reactivity. In a kind of calcium phosphate cement, named brushite, mixing the precursor powders with a citrate compound decreased the reactivity of the powders. The event resulted from the bond appeared between citrate compounds and released calcium ions that prevented nucleation and growth of brushite compound [38]. By adding diammonium citrate $((NH_4)_2C_6H_6O_7)$ to the precursor powder of magnesium-calcium phosphate cement, a negative charge was created on the surface of the powders that prevented agglomeration and consequently improved the cement paste injectability [41]. Mestres and Ginebra [49] evaluated the effect of sodium borate decahydrate (Borax, $Na_2B_4O_7.10H_2O$) on the setting reaction of magnesium phosphate cement. They reported that the formation of $B_4O_7^{2-}$ ions on the surface of magnesium oxide, as a precursor powder, decreased the powder reactivity. In the review study [38], some additives that can improve the injectability of brushite cement by controlling the setting process were mentioned. The additives were classified as follows: (1) pyrophosphate compounds including sodium and calcium pyrophosphate that by affecting the presence of phosphate increases the required energy for nucleation; (2) sulfate compounds such as calcium sulfate dihydrate and sulfuric acid by replacing phosphate ions the controls setting process; and (3) organic acids containing citric acid, glycolic acid, and tartaric acid by binding to calcium ions decreases the setting rate.

The variation of precursor powder reactivity by modifying its composition can affect the mechanical properties of the final cement. For example, Alkhraisat et al. [50] reported that adding magnesium to β-tricalcium phosphate, as a precursor powder of brushite cement, enhanced the chemical stability of the powders and could improve the mechanical properties of brushite cement. As also mentioned in Fig. 1.9, the reactivity of the three different precursor powders changed based on their compositions. $Mg_{2.25}Ca_{0.75}(PO_4)_2$ that possesses lower reactivity compared to other ones can create cement with the highest compressive strength even after 120 min grinding of precursor powders. However, the most reactive precursor powder, $Ca_3(PO_4)_2$ can improve the compressive strength of the final cement when the powder was ground for 30–60 min. Because more grinding time decreased the average size of the precursor powders, their high reactivity caused inappropriate preparation of the cement and in consequence, reduced the compressive strength of the final cement [46]. Esnaashary et al. [51, 52] also reported that by adding calcium and sodium to the precursor powders of magnesium phosphate cement, the reactivity of the powders was reduced, and the setting time and mechanical strength of the cement became more controlled.

In polymerization, one of the practical parameters on reactivity is the amount of initiator. As more initiator is used, the production of free radicals and in consequence formation of crosslinks are accelerated which reduces the gelling time. In addition, a lower amount of initiator can reduce the length of crosslinks and improves the mechanical properties of final structure [44].

1.9   Tissue Engineering

Due to the lifestyle of modern people, the bone disease is a common symptom. As reported, annually 15 million fractures occur that impose a considerable cost to society. As mentioned in the previous sections, gold standard methods such as autograft and allograft for treating bone defects are connected with some restrictions. The restriction of autograft is the requirement for an additional surgical site that can cause the occurrence of morbidity in the donor site and interfering convenience of a patient. The other method may cause infection transmission and immune rejection [53]. To overcome these problems, a new method named bone tissue engineering has been introduced. In this manner, synthetic grafts are fabricated by mimicking natural bone tissue. In this book, applying tissue engineering strategy on bone cement will be evaluated based on three important factors. The first is the scaffold whereas the composition, geometry, and architecture of natural bone tissue is mimicked. The second chapter, entitled conductivity: materials design, would introduce different compositions applied in producing implants and how to fabricate porous structures. The second factor is the cell source established according to different cell lineages that play their roles in bone tissue. The third chapter, named productivity: cell, embeds this subject. The third factor relates to growth mechanisms, including the one that will be explained in the fourth chapter named inductivity: bioactive agents. These three factors made the elements of the tissue engineering pyramid as shown in Fig. 1.10.

Fig. 1.10   The pyramid of tissue engineering

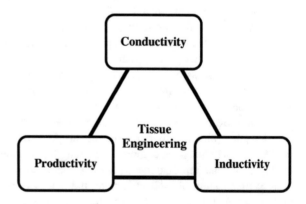

References

1. R.G. Blezard, The history of calcareous cements, in *Lea's Chemistry of Cement and Concrete*, ed. by P. Hewlett, 4th edn. (Butterworth-Heinemann, Oxford, 2004), pp. 1–23
2. https://www.flickr.com/photos/stanrandom/94221045/
3. J.H. Sharp, Surely we know all about cement—don't we? Adv. Appl. Ceram. **105**, 162–174 (2006). https://doi.org/10.1179/174367606X115904
4. M.M. Masoumi, H. Banakar, B. Boroomand, Review of an ancient persian lime mortar "Sarooj". Malaysian J. Civ. Eng. **27**, 94–109 (2015). https://doi.org/10.11113/mjce.v27n1.361
5. R. Eires, A. Camões, S. Jalali, Earth architecture: ancient and new methods for durability improvement, in *Structures and Architecture: New Concepts, Applications and Challenges*, ed. by P.J. Cruz (CRC Press, London, 2013), pp. 962–970
6. No Title. https://en.wikipedia.org/wiki/Sarooj. Accessed 25 Sept 2019
7. F. Chen, X. Ma, Y. Yu, C. Liu, Calcium phosphate bone cements: their development and clinical applications, in *Developments and Applications of Calcium Phosphate Bone Cements*, ed. by C. Liu, H. He (Springer, Singapore, 2018), pp. 1–39
8. R. Vaishya, M. Chauhan, A. Vaish, Bone cement. J. Clin. Orthop. Trauma **4**, 157–163 (2013). https://doi.org/10.1016/j.jcot.2013.11.005
9. M.C. Birt, D.W. Anderson, E.B. Toby, J. Wang, Osteomyelitis: recent advances in pathophysiology and therapeutic strategies. J. Orthop. **14**, 45–52 (2017). https://doi.org/10.1016/j.jor.2016.10.004
10. B. Maurel, T. Le Corroller, G. Bierry, X. Buy, P. Host, A. Gangi, Treatment of symptomatic para-articular intraosseous cysts by percutaneous injection of bone cement. Skelet. Radiol. **42**, 43–48 (2013). https://doi.org/10.1007/s00256-012-1392-7
11. Z. He, Q. Zhai, M. Hu, C. Cao, J. Wang, H. Yang, B. Li, Bone cements for percutaneous vertebroplasty and balloon kyphoplasty: current status and future developments. J. Orthop. Transl. **3**, 1–11 (2015). https://doi.org/10.1016/j.jot.2014.11.002
12. L.E. Mermelstein, R.F. McLain, S.A. Yerby, Reinforcement of thoracolumbar burst fractures with calcium phosphate cement. Spine (Phila Pa 1976) **23**, 664–670 (1998). https://doi.org/10.1097/00007632-199803150-00004
13. A. Vaccaro, F. Kandziora, M. Fehlings, R. Shanmughanathan, MIS posterior short segment fixation with pedicle screws, in *AO Fund* (2014). https://www2.aofoundation.org/wps/portal/surgery?showPage=redfix&bone=Spine&segment=TraumaThoracolumbar&classification=53-A4&treatment=&method=MISposteriorshortsegmentfixationwithpediclescrews&implantstype=&approach=&redfix_url=1417446919819&Langua
14. S. Rupprecht, H.A. Merten, P. Kessler, J. Wiltfang, Hydroxyapatite cement (BoneSource™) for repair of critical sized calvarian defects—an experimental study. J. Cranio-Maxillofacial Surg. **31**, 149–153 (2003). https://doi.org/10.1016/S1010-5182(03)00017-9
15. G. Lewis, Properties of acrylic bone cement: state of the art review. J. Biomed. Mater. Res. **38**, 155–182 (1997). https://doi.org/10.1002/(SICI)1097-4636(199722)38:2%3c155:AID-JBM10%3e3.0.CO;2-C
16. C.V. Rahman, A. Saeed, L.J. White, T.W.A. Gould, G.T.S. Kirby, M.J. Sawkins, C. Alexander, F.R.A.J. Rose, K.M. Shakesheff, Chemistry of polymer and ceramic-based injectable scaffolds and their applications in regenerative medicine. Chem. Mater. **24**, 781–795 (2012). https://doi.org/10.1021/cm202708n
17. K.L. Low, S.H. Tan, S.H.S. Zein, J.A. Roether, V. Mouriño, A.R. Boccaccini, Calcium phosphate-based composites as injectable bone substitute materials. J. Biomed. Mater. Res. B Appl. Biomater. **94**, 273–286 (2010). https://doi.org/10.1002/jbm.b.31619
18. M.P. Ginebra, M. Espanol, E.B. Montufar, R.A. Perez, G. Mestres, New processing approaches in calcium phosphate cements and their applications in regenerative medicine. Acta Biomater. **6**, 2863–2873 (2010). https://doi.org/10.1016/j.actbio.2010.01.036
19. W.C. Dewey, Arrhenius relationships from the molecule and cell to the clinic. Int. J. Hyperth. **25**, 3–20 (2009). https://doi.org/10.1080/02656730902747919

20. S.B. Field, C.C. Morris, The relationship between heating time and temperature: its relevance to clinical hyperthermia. Radiother. Oncol. **1**, 179–186 (1983). https://doi.org/10.1016/S0167-8140(83)80020-6

21. G. Kraaij, D.F. Malan, H.J.L. van der Heide, J. Dankelman, R.G.H.H. Nelissen, E.R. Valstar, Comparison of Ho:YAG laser and coblation for interface tissue removal in minimally invasive hip refixation procedures. Med. Eng. Phys. **34**, 370–377 (2012). https://doi.org/10.1016/j.medengphy.2011.07.029

22. A.R. Eriksson, T. Albrektsson, Temperature threshold levels for heat-induced bone tissue injury: a vital-microscopic study in the rabbit. J. Prosthet. Dent. **50**, 101–107 (1983). https://doi.org/10.1016/0022-3913(83)90174-9

23. H. Deramond, N. Wright, S. Belkoff, Temperature elevation caused by bone cement polymerization during vertebroplasty. Bone **25**, 17S–21S (1999). https://doi.org/10.1016/S8756-3282(99)00127-1

24. J.K. Hurley, Acid-base balance: normal regulation and clinical application. Curr. Probl. Pediatr. **9**, 1–43 (1979). https://doi.org/10.1016/S0045-9380(79)80014-6

25. OpenStax, Fluid, electrolyte, and acid-base balance, in *Anatomy and Physiology* (OpenStax CNX, 2016)

26. L.L. Hamm, N. Nakhoul, K.S. Hering-Smith, Acid-base homeostasis. Clin. J. Am. Soc. Nephrol. **10**, 2232–2242 (2015). https://doi.org/10.2215/CJN.07400715

27. F.C.M. Driessens, J.A. Planell, M.G. Boltong, I. Khairoun, M.P. Ginebra, Osteotransductive bone cements. Proc. Inst. Mech. Eng. Part H J. Eng. Med. **212**, 427–435 (1998). https://doi.org/10.1243/0954411981534196

28. S.V. Dorozhkin, Self-setting calcium orthophosphate formulations. J. Funct. Biomater. **4**, 209–311 (2013). https://doi.org/10.3390/jfb4040209

29. R. Nadiv, G. Vasilyev, M. Shtein, A. Peled, E. Zussman, O. Regev, The multiple roles of a dispersant in nanocomposite systems. Compos Sci. Technol. **133**, 192–199 (2016). https://doi.org/10.1016/j.compscitech.2016.07.008

30. M. Bohner, G. Baroud, Injectability of calcium phosphate pastes. Biomaterials **26**, 1553–1563 (2005). https://doi.org/10.1016/j.biomaterials.2004.05.010

31. J.A.J. Wagoner, B.A. Herschler, A review of the mechanical behavior of CaP and CaP/polymer composites for applications in bone replacement and repair. Acta Biomater. **7**, 16–30 (2011). https://doi.org/10.1016/j.actbio.2010.07.012

32. A. Sugawara, K. Asaoka, S.J. Ding, Calcium phosphate-based cements: clinical needs and recent progress. J. Mater. Chem. B **1**, 1081–1089 (2013). https://doi.org/10.1039/c2tb00061j

33. F. Chen, C. Liu, J. Wei, X. Chen, Physicochemical properties and biocompatibility of white dextrin modified injectable calcium-magnesium phosphate cement. Int. J. Appl. Ceram. Technol. **9**, 979–990 (2012). https://doi.org/10.1111/j.1744-7402.2011.02705.x

34. J.W. Nicholson, *The Chemistry of Polymers*, 3rd edn. (RSC Publishing, Dorset, 2006)

35. A.S. Wagh, *Chemically Bonded Phosphate Ceramics: Twenty-First Century Materials with Diverse Applications* (Elsevier Ltd, Oxford, 2004)

36. J.W. Bullard, H.M. Jennings, R.A. Livingston, A. Nonat, G.W. Scherer, J.S. Schweitzer, K.L. Scrivener, J.J. Thomas, Mechanisms of cement hydration. Cem. Concr. Res. **41**, 1208–1223 (2011). https://doi.org/10.1016/j.cemconres.2010.09.011

37. J.J. Thomas, J.J. Biernacki, J.W. Bullard, S. Bishnoi, J.S. Dolado, G.W. Scherer, A. Luttge, Modeling and simulation of cement hydration kinetics and microstructure development. Cem. Concr. Res. **41**, 1257–1278 (2011). https://doi.org/10.1016/j.cemconres.2010.10.004

38. F. Tamimi, Z. Sheikh, J. Barralet, Dicalcium phosphate cements: brushite and monetite. Acta Biomater. **8**, 474–487 (2012). https://doi.org/10.1016/j.actbio.2011.08.005

39. F. Wu, J. Su, J. Wei, H. Guo, C. Liu, Injectable bioactive calcium-magnesium phosphate cement for bone regeneration. Biomed. Mater. **3**, 044105 (2008). https://doi.org/10.1088/1748-6041/3/4/044105

40. M. Habib, G. Baroud, F. Gitzhofer, M. Bohner, Mechanisms underlying the limited injectability of hydraulic calcium phosphate paste. Acta Biomater. **4**, 1465–1471 (2008). https://doi.org/10.1016/j.actbio.2008.03.004

41. C. Moseke, V. Saratsis, U. Gbureck, Injectability and mechanical properties of magnesium phosphate cements. J. Mater. Sci. Mater. Med. **22**, 2591–2598 (2011). https://doi.org/10.1007/s10856-011-4442-0

42. M. Espanol, R.A. Perez, E.B. Montufar, C. Marichal, A. Sacco, M.P. Ginebra, Intrinsic porosity of calcium phosphate cements and its significance for drug delivery and tissue engineering applications. Acta Biomater. **5**, 2752–2762 (2009). https://doi.org/10.1016/j.actbio.2009.03.011

43. H.A. Samad, M. Jaafar, Effect of polymethyl methacrylate (PMMA) powder to liquid monomer (P/L) ratio and powder molecular weight on the properties of PMMA cement. Polym. Plast. Technol. Eng. **48**, 554–560 (2009). https://doi.org/10.1080/03602550902824374

44. S.J. Peter, Injectable, in situ polymerizable, biodegradable scaffold based on poly(propylene fumarate) for guided bone regeneration, Rice University, 1998

45. E.B. Montufar, Y. Maazouz, M.P. Ginebra, Relevance of the setting reaction to the injectability of tricalcium phosphate pastes. Acta Biomater. **9**, 6188–6198 (2013). https://doi.org/10.1016/j.actbio.2012.11.028

46. U. Klammert, T. Reuther, M. Blank, I. Reske, J.E. Barralet, L.M. Grover, A.C. Kübler, U. Gbureck, Phase composition, mechanical performance and in vitro biocompatibility of hydraulic setting calcium magnesium phosphate cement. Acta Biomater. **6**, 1529–1535 (2010). https://doi.org/10.1016/j.actbio.2009.10.021

47. E.B. Montufar, T. Traykova, E. Schacht, L. Ambrosio, M. Santin, J.A. Planell, M.-P. Ginebra, Self-hardening calcium deficient hydroxyapatite/gelatine foams for bone regeneration. J. Mater. Sci. Mater. Med. **21**, 863–869 (2010). https://doi.org/10.1007/s10856-009-3918-7

48. F. Perut, E.B. Montufar, G. Ciapetti, M. Santin, J. Salvage, T. Traykova, J.A. Planell, M.P. Ginebra, N. Baldini, Novel soybean/gelatine-based bioactive and injectable hydroxyapatite foam: material properties and cell response. Acta Biomater. **7**, 1780–1787 (2011). https://doi.org/10.1016/j.actbio.2010.12.012

49. G. Mestres, M.-P. Ginebra, Novel magnesium phosphate cements with high early strength and antibacterial properties. Acta Biomater. **7**, 1853–1861 (2011). https://doi.org/10.1016/j.actbio.2010.12.008

50. M.H. Alkhraisat, J. Cabrejos-Azama, C.R. Rodríguez, L.B. Jerez, E.L. Cabarcos, Magnesium substitution in brushite cements. Mater. Sci. Eng. C **33**, 475–481 (2013). https://doi.org/10.1016/j.msec.2012.09.017

51. M.H. Esnaashary, H.R. Rezaie, A. Khavandi, J. Javadpour, Solubility controlling of the precursor powders of magnesium phosphate cement by changing the powder composition. Adv. Appl. Ceram. **116**, 286–292 (2017). https://doi.org/10.1080/17436753.2017.1315860

52. M.H. Esnaashary, H.R. Rezaie, A. Khavandi, J. Javadpour, Evaluation of setting time and compressive strength of a new bone cement precursor powder containing Mg–Na–Ca. Proc. Inst. Mech. Eng. Part H J. Eng. Med. **232**, 1017–1024 (2018). https://doi.org/10.1177/0954411918796048

53. K. Shin, T. Acri, S. Geary, A.K. Salem, Biomimetic mineralization of biomaterials using simulated body fluids for bone tissue engineering and regenerative medicine. Tissue Eng. Part A **23**, 1169–1180 (2017). https://doi.org/10.1089/ten.tea.2016.0556

Chapter 2
Conductivity: Materials Design

2.1 Introduction

According to Hench's definition, osteoconductivity appears by conforming newly grown bone with the free surface of the implant. The feature is very crucial for the resistance of the implant for a long time. Bone inherently cannot cure by its own. Before its growth, vessels, fibrovascular tissues, and osteoprogenitor cells should grow on an implant. In this manner, three critical properties should be involved to an implant: (1) Surface chemistry: the compound that mimics composition of natural bone tissue, including calcium phosphate, magnesium phosphate, gelatin, collagen, bioactive glass, and polypropylene fumarate; (2) Surface topography: surface roughness that can improve the interaction of cell-implant; (3) Architectural geometry: fibril and porous structure of bone extracellular matrix [1]. To understand these features, the bone structure is introduced in the next section. Based on the composition of bone, the conventional composition of bone cement is studied. Then by evaluating different production methods of porous structures, it is discussed how to apply the bone architecture on the bone cement.

2.2 Bone

Bone is a natural nanocomposite that occupies many crucial roles such as mechanically stabilizing body, preserving sensitive inner organs, the reservoir of requirement ions including calcium, magnesium, sodium, and phosphor, and balancing the concentration of blood ions [2]. According to Fig. 2.1, the bone structure consists of two spongy and compact parts. The former embeds bone marrow. The compact part is built from tubular structures named osteon whereat Haversian canals go through them. Osteon forms of a collection of collagenous fibril and nano-crystals of hydroxyapatite coat them [3]. Generally, about 20 wt% of bone is composed of collagen, and 70 wt% is composed of hydroxyapatite [4]. The mentioned structures fill the

H. Reza Rezaie et al., *Bone Cement*, SpringerBriefs in Applied
Sciences and Technology, https://doi.org/10.1007/978-3-030-39716-6_2

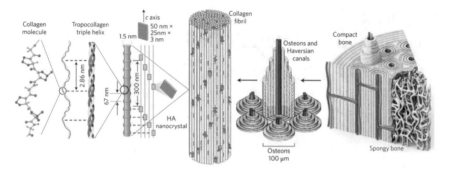

Fig. 2.1 The hierarchical structure of bone, from macro to nano-scale (Reprinted from [3] by permission from Springer Nature)

intra-space of bone cells, which is called the extracellular matrix. The composition of the extracellular matrix depends on the type of surrounding cells, and also the composition of the extracellular matrix determines the cell types that can grow on them. In other words, there is a dynamic balance between the extracellular matrix and cell type [5, 6].

2.3 Bone Cement Compositions

According to bone composition or demanded properties, different compositions of bone cement have been introduced. In the following sections, some polymer and ceramic compositions including polymethyl methacrylate, polypropylene fumarate, calcium phosphate cement, and magnesium phosphate cement are considered.

2.3.1 Polymethyl Methacrylate

An acrylic polymer named polymethyl methacrylate has been introduced as primary bone cement. The cement consists of two phases: a liquid phase includes monomers of methyl methacrylate, and a solid phase includes pre-polymerized powder of either pure polymethyl methacrylate or polymethyl methacrylate accompanied by another polymer as a copolymer. By mixing the two phases, during the polymerization process, the monomers start to accumulate around the pre-polymerized powders and produce a rigid body [7]. Like any polymer, polymerization of polymethyl methacrylate follows three main stages, including initiation, propagation, and termination. The schematic of its reaction in the presence of 2,2′-azo-bis-isobutyrylnitrile as an initiator is shown in Fig. 2.2 [8].

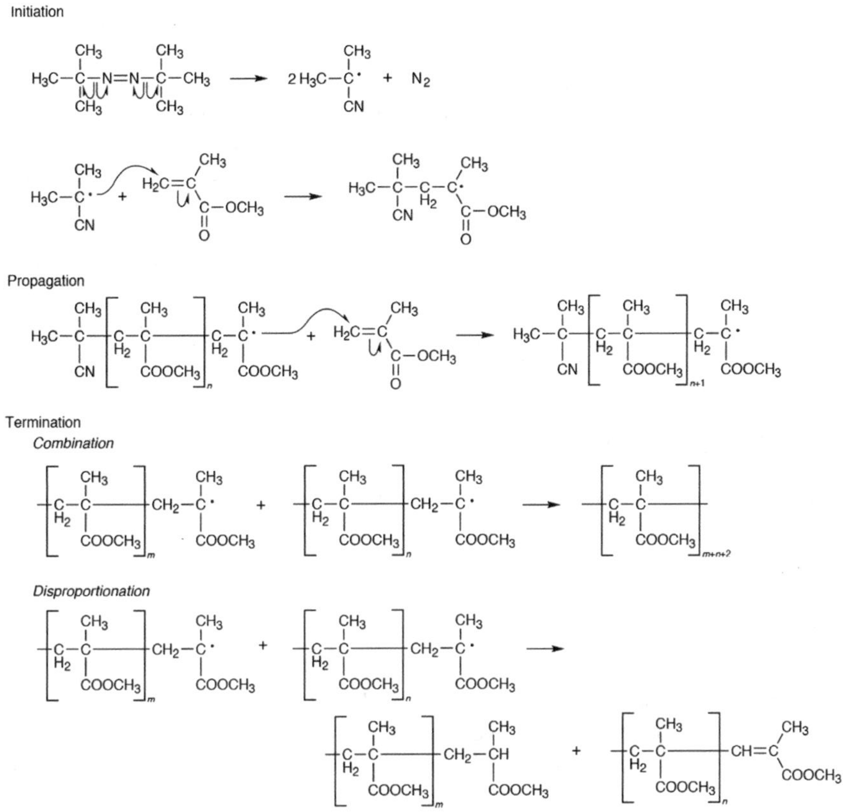

Initiation

Propagation

Termination

Combination

Disproportionation

Fig. 2.2 Polymerization of methyl methacrylate in the presence of 2 2′-azobis-isobutyronitrile (Reprinted from [8] by permission from ACS Publication)

Other materials are also combined with the mixture. One of them is hydroquinone that prevents maturation of the monomers by light or temperature. Di-benzoyl peroxide as an initiator and N,N-dimethyl-p-toluidine as an accelerator are mixed by the phases to control the polymerization process. Because the evaluation of the treated part of the human body is vital for surgeons and the polymer is transparent relates to evaluation techniques, such as CT-scan, some additives including zirconium dioxide or barium sulfate are added to make the polymer radiopaque. Some antibiotics to sterilize the treated site are also incorporated into the compound consisting of Gentamycin, Tobra- mycin, Erythromycin, Cefuroxime, Vancomycin, and Colistin [7].

Some drawbacks faced the application of polymethyl methacrylate such as (1) exothermic polymerization reaction of the polymer that can cause thermal necrosis of surrounding tissues and local interfere in the blood circulation (the exothermic reaction originated from cleavage of a carbon double bond to a single bond that produces 130 cal/g of heat for each bond [9]), (2) release of unreacted monomers that are cytotoxic and cause chemical necrosis, (3) a significant mismatch between the

Table 2.1 Commercial polymethyl methacrylate cement (Adapted from [16] by permission from Elsevier)

Brand	Working time (min)	Setting time (min)	Viscosity	Bending modulus (MPa)	Bending strength (MPa)	Compressive strength (MPa)	Supplier
CMW1	6.5	11	High	2634	67.81	94.4	Depuy
Smartset HV	8	12.5	High	3010	64.32	86.54	Depuy
Endurance	8	14	Low	2896	76.1	94	Depuy
Smartset MV	8	14	Medium	3010	64.32	70	Depuy
Simplex P	7	14.3	Medium	2681	71	90.32	Stryker
Spineplex	10–12	8.7	Low	–	55.1	80.91	Stryker
Palacos R	5	12.5	High	2628	72.2	79.6	Heraeus
Osteopal V	8	14	Low	3504	47	82	Heraeus
Cobalt HV	5	10	High	–	67.84	96.04	Biomet
Osteobond	5	14.5	Low	2828	73.7	104.6	Zimmer
KyphX HV-R	8	20	High	–	–	111	Kyphon
ABC	4.5–6.5	12	Medium	3300	68	93	Tianjin Ins.

stiffness of cement and bone, (4) non-degradability of the polymer in body condition [10]. Many kinds of research have been done to overcome the mentioned shortages. Strontium-hydroxyapatite was added to the polymer to improve its radiopacity and bioactivity [11]. In another research, Wang et al. [12] reported that hydroxyapatite could enhance the mechanical properties and bioactivity of the polymer. Combining $Ca_2MgSi_2O_7$ with polymethyl methacrylate improved the bioactivity and reduced temperature of the exothermic reaction [13]. Marrs et al. [14] enhanced the fatigue strength of the polymer by reinforcing it with carbon nanotubes. Polyethylene glycol as a phase change material was mixed with polymethyl methacrylate to adsorb heat produced in the polymerization process. The melting point of the polyethylene glycol can be modified by its molecular weight, and so its heat absorption can be tailored [15]. Information of some of the commercialized cement, such as their brands, manufacturers, viscosity types, and markets, are tabulated [16] in Table 2.1.

2.3.2 Polypropylene Fumarate

Polypropylene fumarate is an unsaturated linear polyester that possesses two attractive features: (1) its carbon double bond in the fumaric acid unit allows crosslinking of the two polymer chains via covalent bonds, as shown in Fig. 2.3; (2) degradation of the polymer via hydrolysis of ester linkages to two biocompatible units including fumaric acid and propylene glycol, as shown in Fig. 2.3 [17, 18]. Compared to

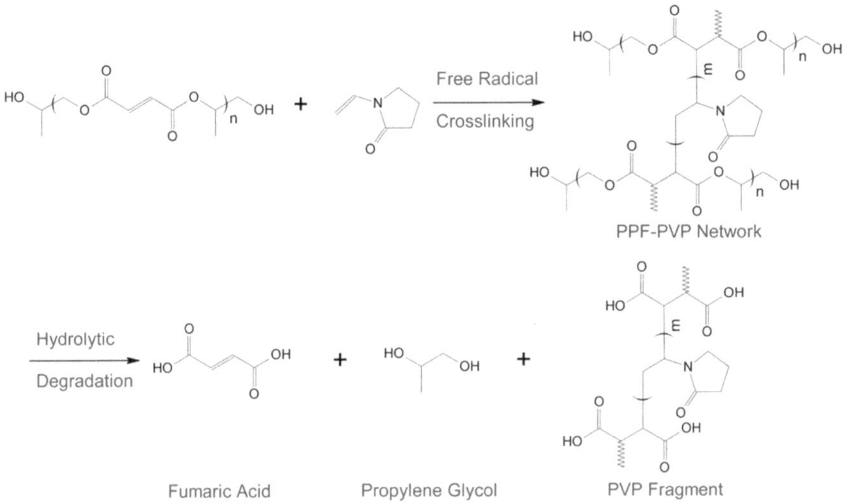

Fig. 2.3 Polymerization of polypropylene fumarate in the presence of n-vinyl pyrrolidine and its degradation to fumaric acid and propylene glycol (Reprinted from [17] by permission from RSC)

polymethyl methacrylate, polypropylene fumarate releases a lower amount of heat during polymerization [19]. Among different used crosslinking agents, it can be mentioned to n-vinyl pyrrolidine [20], polyethylene glycol-dimethacrylate [21], and diethyl fumarate [22]. Many kinds of research have attempted to improve its mechanical and biological properties of polypropylene fumarate by adding hydroxyapatite [23], β-tricalcium phosphate [19], carbon nanotube [24, 25], and titanium oxide [17].

2.3.3 Calcium Phosphate Cement

Usually, calcium phosphate cement is available in the two categories, based on its final composition, including apatite cement and brushite cement. Apatite cement, generally known with the formula of $Ca_{10}(PO_4)_6(OH)_2$, is produced from hydrolysis of α-tricalcium phosphate, α-$Ca_3(PO_4)_2$, or acid-base reaction between tetracalcium phosphate, $Ca_4(PO_4)_2O$, with monocalcium monohydrate, $Ca(H_2PO_4)_2 \cdot H_2O$, or dicalcium phosphate dihydrate, $CaHPO_4 \cdot 2H_2O$. The common reactions are as follows [26, 27]:

$$5Ca_3(PO_4)_2 + 3H_2O \rightarrow 3Ca_5(PO_4)_3OH + H_3PO_4$$

$$3\alpha - Ca_3(PO_4)_2 + H_2O \rightarrow Ca_9(HPO_4)(PO_4)_5OH$$

$$3Ca_4(PO_4)_2O + 3H_2O \rightarrow 2Ca_5(PO_4)_3OH + 2Ca(OH)_2$$

$$7Ca_4(PO_4)_2O + 2Ca(H_2PO_4)_2 \cdot H_2O \rightarrow 6Ca_5(PO_4)_3OH + 3H_2O$$

$$2Ca_4(PO_4)_2O + Ca(H_2PO_4)_2 \cdot H_2O \rightarrow Ca_9(HPO_4)(PO_4)_5OH + 2H_2O$$

$$Ca_4(PO_4)_2O + CaHPO_4 \cdot 2H_2O \rightarrow Ca_5(PO_4)_3OH + 2H_2O$$

$$3Ca_4(PO_4)_2O + 6CaHPO_4 \cdot 2H_2O \rightarrow 2Ca_9(HPO_4)(PO_4)_5OH + 13H_2O$$

$$Ca_4(PO_4)_2O + CaHPO_4 \rightarrow Ca_5(PO_4)_3OH$$

$$3Ca_4(PO_4)_2O + 6CaHPO_4 \rightarrow 2Ca_9(HPO_4)(PO_4)_5OH + H_2O$$

$$3Ca_4(PO_4)_2O + Ca_8H_2(PO_4)_6 \cdot 5H_2O \rightarrow 4Ca_5(PO_4)_3OH + 4H_2O$$

$$3Ca_4(PO_4)_2O + 3Ca_8H_2(PO_4)_6 \cdot 5H_2O \rightarrow 4Ca_9(HPO_4)(PO_4)_5OH + 14H_2O$$

$$Ca_4(PO_4)_2O + 2Ca_3(PO_4)_2 + H_2O \rightarrow Ca_5(PO_4)_3OH$$

These reactions occur at pH greater than 5 [28]. Ewald et al. [28] reported the non-cytotoxicity effect of apatite cement and its ability to enhance the proliferation and adhesion of bone cells. In addition, because the body fluids are supersaturated with apatite composition and stability of the compound is at the pH of the body fluids, apatite cement cannot chemically dissolve. The cement only can dissolve based on the activity of the surrounded cell; hence, its dissolution rate is very weak [29]. As Fig. 2.4 shows, after implanting apatite cement for 15 months, its structure did not deform significantly and approximately preserved its original dimensions [30].

Brushite, dicalcium phosphate dihydrate $CaHPO_4 \cdot 2H_2O$, is another kind of calcium phosphate cement that is produced from a reaction between β-tricalcium phosphate and monocalcium phosphate monohydrate [31]. The common reaction to produce brushite cement is as follows:

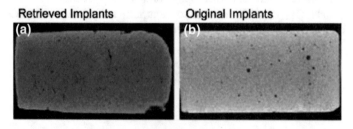

Fig. 2.4 μ-CT evaluation of implanted apatite cement at **a** implanting time and **b** after 15 months (Reprinted from [30] by permission from Elsevier)

Table 2.2 Commercial calcium phosphate cement (Adapted from [16] by permission from Elsevier)

Brand	Initial setting time (min)	Hardening time (h)	End product	Porosity (%)	Pore size (um)	Strength (MPa)	Supplier
BoneSource	7	4	Apatite	5–10	33	26	Stryker
Norian SRS	10–15	12	Carbonated Apatite	50	47	50	Synthes
α-BSM	15–20	1	Apatite	80	<1	4	ETEX
Biopex	7–10	24	Apatite	40–50	–	80	Mitsubishi
Calcibon	10	6	Carbonated Apatite	30–40	42	60	Biomet–Merck
Cementek	3–15	24–48	Apatite	50	26	13	Teknimed
Graftys HBS	15	72	Apatite	65–70	100–300	12	Graftys
Graftys Quickset	8	24	Apatite	70	10–100	24	Graftys
Ostim	20	–	Apatite	53	70	0.24	Hereaus
chronOS Inject	612	24	Brushite	60–75	70–170	3	Synthes
Eurobone	3–15	–	Apatite	2	162	17	FH Orthopedics
KyphOs FS	5	24	Apatite	–	–	61	Kyphon

$$\beta - Ca_3(PO_4)_2 + Ca(H_2PO_4)_2 \cdot H_2O + 7H_2O \rightarrow 4CaHPO_4 \cdot 2H_2O$$

Despite the apatite type, brushite precipitated at pH lower than 5 [28]. In addition, the cement creates with a very high rate and also rapidly dissolves in biological fluids [31, 32]. However, brushite is degradable, it can transform to a more stable composition in body fluids which is apatite [33]. Like other calcium phosphates, brushite did not show any cytotoxic behavior and improves the proliferation and adhesion of bone cells [28, 34].

Many kinds of calcium phosphate cement are commercialized that can be found in Table 2.2.

2.3.4 Magnesium Phosphate Cement

Recently, a new kind of ceramic cement, named magnesium phosphate cement, was introduced. The cement is produced in a pH higher than 5 by the reaction of magnesium oxide (MgO) or trimagnesium phosphate ($Mg_3(PO_4)_2$) with diammonium hydrogen phosphate (($NH_4)_2HPO_4$), ammonium dihydrogen phosphate ($NH_4H_2PO_4$) or potassium dihydrogen phosphate (KH_2PO_4). The final composition of this cement is struvite ($MgNH_4PO_4 \cdot 6H_2O$), newberyite ($MgHPO_4 \cdot 3H_2O$),

K-struvite ($KMgPO_4 \cdot 6H_2O$), and hannayite ($Mg_3(NH_4)_2H_4(PO_4)_4 \cdot 8H_2O$) [35–37]. The common reactions are as follows [38]:

$$MgO + H_3PO_4 + 2H_2O \rightarrow MgHPO_4 \cdot 3H_2O$$

$$MgO + NH_4H_2PO_4 + 5H_2O \rightarrow MgNH_4PO_4 \cdot 6H_2O$$

$$MgO + (NH_4)_2HPO_4 + 5H_2O \rightarrow MgNH_4PO_4 \cdot 6H_2O + NH_3 \uparrow$$

$$MgO + KH_2PO_4 + 5H_2O \rightarrow MgKPO_4 \cdot 6H_2O$$

$$2Mg_3(PO_4)_2 + 3(NH_4)_2HPO_4 + 36H_2O \rightarrow 6MgNH_4PO_4 \cdot 6H_2O + H_3PO_4$$

$$Mg_3(PO_4)_2 + (NH_4)_2HPO_4 + 15H_2O \rightarrow 2MgNH_4PO_4 \cdot 6H_2O + MgHPO_4 \cdot 3H_2O$$

$$Mg_3(PO_4)_2 + NH_4H_2PO_4/(NH_4)_2HPO_4 \rightarrow Mg_3(NH_4)_2H_4(PO_4)_4 \cdot 8H_2O$$

The high adsorbed water in the structure of this cement reduces the sensitivity of powder to liquid ratio on the final mechanical properties of the cement. This feature enables higher control on injectability and setting process of the cement [36]. In vitro evaluations of the cement reported its acceptable biocompatibility and suitable activity of the osteoblasts and osteoclasts on its surface [39, 40]. The cement, like brushite, is biodegradable either chemically or by the activity of osteoclasts [29].

2.3.5 Other Types of Cement

In addition to the mentioned compositions, some compositions such as zinc phosphate, zinc polycarboxylate, and glass polyalkenoate are used as dental/bone cement. To understand in more detail about the cement, one can read the review written by Kenny and Buggy [41].

2.4 Scaffold

Bone cement can be hardened at room temperature. This feature makes possible mimicking bone structure, by incorporating porosity, bioactive agents, and even cells in a cement structure. In this order, produce scaffold from bone cement has been considered. In the following sections, different methods for fabricating scaffold and some examples of applying the methods on bone cement are addressed.

2.4.1 Properties

Despite common properties expected from biomaterials including biocompatibility, biodegradability, osteoconductivity, and osteoinductivity, other features are required for a structure to be a scaffold:

(1) The structure should consist of a porosity with a volume content of more than 90% and with diameters varying around 300–500 µm. The porous structure should be interconnected. These features allow surrounding tissue, including cells and required nutrients, to penetrate in the structure;

(2) The structure should possess suitable mechanical properties to bear applied load until it is replaced by new formed extracellular matrix;

(3) It should be possible to produce a porous structure with desired properties such as its apparent shape and pore shape;

(4) Its production process should be cost-effective [42].

2.4.2 Production Methods

Many production techniques have been introduced to fabricate a scaffold. In this section, some of them are briefly described, including freeze-drying, particulate leaching, gas foaming, electrospinning, melt molding, phase separation, fiber bonding, and rapid prototyping.

2.4.2.1 Freeze-Drying

In the freeze-drying technique, a polymer/ceramic is dissolved in water or an organic solvent and then emulsified in water. The emulsion consists of a continuous polymer/ceramic solution with dispersed water sections in it. By freezing the emulsion and reducing the environmental pressure, water and solvent would be sublimed.

Advantages: A single-step process without requirement for a separate leaching step, no need to heating, and producing highly porous scaffold with interconnected structure.

Disadvantages: Diameter of pores is small, the process requires a long time, and usage solvent is usually toxic [43, 44].

2.4.2.2 Particulate Leaching

The particulate leaching method is employed in two approaches: solvent casting and melt molding. The former is done by dissolving a polymer in a solvent and mixing it with porogen agents. The mixture is molded, let the solvent to evaporate, and then the porogen agents are eliminated by washing with water. In the second method, a polymer is blended with porogen agents and heated. The porogen agents are removed by water.

Advantages: Producing a highly porous structure with large-diameter pores, the ability to control the pore size, and its simplicity.

Disadvantages: Internal structure of scaffold varies from the surface contacted by the mold to the surface contacted by air, and remaining some porogen agents especially in the central region of the scaffold [43, 44].

2.4.2.3 Gas Foaming

In the gas foaming method, polymer pellets are compressed and heated. High pressure of carbon dioxide is exerted on the obtained molded polymer for 3 days. After saturation of the polymer by the gas, the pressure is gradually reduced, and a highly porous scaffold is obtained. Another method is also common by mixing ammonium bicarbonate with a polymer solution. During polymerization, the chemical reaction of the porogen agents produces gas and leaves a porous structure.

Advantages: In the first method, toxic solvents are not required, the second method avoids using high temperature, and obtains a more uniform pore structure compared to the particulate leaching method.

Disadvantages: Formation of a polymer skin on the scaffold surface and inability to produce complex scaffold shapes [43, 44].

2.4.2.4 Electrospinning

Electrospinning is an old approach originated from electrostatic spraying. Through this method, the polymer solution/melt is placed among an electric potential difference applied to a needle containing the solution/melt and a grounded gathering screen/plate/mandrel. The solution is to drown out in a shape of a filament, and in its rout, its contained solvent evaporates and remained fibers, in the range of 50 nm to 10 μm, are collected on the plate. Finally, a textile of randomly oriented fibers with residual interconnected pore structure among the fibers is obtained.

Advantages: A simple method that in the best approach can mimic the extracellular matrix.

Disadvantages: Lack of mechanical properties and using toxic solvents [43, 44].

2.4.2.5 Melt Molding

In the melt molding method, polymer particles accompanying microparticles of porogen agents are compacted in a mold and heated. By heating, polymer particles melt, fuse and produce a matrix of the polymer embedded the porogen agents. After cooling, the obtained structure is immersed in water to leach the porogen agents.

Advantages: Avoid using toxic solvent and the ability to control pore geometry.

Disadvantages: Restricted to the shape of the mold and remaining some porogen agents [43, 44].

2.4.2.6 Phase Separation

The phase separation technique is performed in two types: solid-liquid and liquid-liquid. In this method, a polymer is dissolved in a solvent and then quenched into a mold. In this regard, two separate phases, including polymer-rich and polymer-poor phases, are obtained. By applying vacuum drying, the solvent is eliminated and a porous scaffold remains.

Advantages: Creating a uniform pore structure in the range of 50–100 μm, the obtained scaffold can be applied as bioagents delivery vehicle, and it can be combined with other methods of the scaffold fabrication.

Disadvantages: Using toxic solvents [43, 44].

2.4.2.7 Fiber Bonding

In the fiber bonding technique, a distinct bundle of polymer fibers is coated by a solution of a polymer and its solvent. The solvent is removed from the composite by using a vacuum drying method for a day. Then, the remained structure is heated until the fibers bond at their intersection. Again, the coating polymer is dissolved in its solvent and removed via vacuum drying.

Advantages: Creating highly porous and interconnected structure with a large surface area-to-volume ratio.

Disadvantages: Inability to control the pore size, using toxic solvent, restricted to a few numbers of polymers, and inability to apply bioactive agents [43, 44].

2.4.2.8 Rapid Prototyping

In the rapid prototyping techniques, the structure of a demanded scaffold is determined by imaging methods, such as magnetic resonance imaging and computed tomography, and by using computer-aided design, a three-dimensional structure is produced layer by layer. The technique consists of many other methods including stereolithography, laminated object manufacturing, selective laser sintering, fused deposition modeling, solid ground curing, and three-dimensional printing. In the stereolithography method, usually, monomers that are curable by a laser beam are used. Each layer is built based on exposing a laser beam to predetermined sites, and the procedure is repeated for each layer. In the laser selective sintering model, heat-fusible powders are spread on a surface and sintered by applying a laser beam. The produced part is lowered, and again, the powders are spread on and making the next layer. In the fused deposition type, a polymer fiber is extruded from a mobile nozzle and produces a demanded structure layer by layer. In three-dimensional printing, a layer of powder is spread on a surface and an ink containing a binder compound is dropped on the predetermined sites. The powders exposed by the binder are dissolved and join to their neighbor powders. The process is repeated for each layer to complete a demanded structure.

Advantages: Mechanical, physical, and chemical properties of the scaffold can be precisely controlled, and the inner and outer structure of scaffold can be tailored based on its real shape.

Disadvantages: Limited resolution to 50–300 μm and using toxic solvents [43, 44].

2.4.3 Bone Cement Scaffolds

In the next sections, some examples about applying scaffold production methods on various bone cement compositions are addressed.

2.4.3.1 Polymethyl Methacrylate

Shi et al. [45] incorporated gelatin microspheres in polymethyl methacrylate cement to achieve two principal aims. The first was to release gradually a drug encountered in gelatin microspheres and the second was to use the microspheres as porogen agents to gradually produce porosity in the structure of the cement. As shown in Fig. 2.5, in the best-evaluated condition, they can obtain a structure with around 30% porosity.

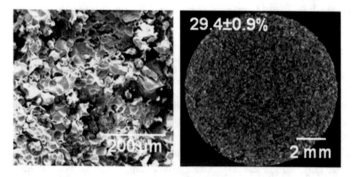

Fig. 2.5 The micrograph and micro-computed tomography image of final porous polymethyl methacrylate cement (Reprinted from [45] by permission from Elsevier)

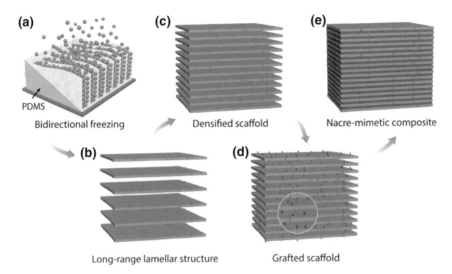

(a) PDMS
Bidirectional freezing

(b) Long-range lamellar structure

(c) Densified scaffold

(d) Grafted scaffold

(e) Nacre-mimetic composite

Fig. 2.6 The schematic of the step-wise procedure of producing hydroxyapatite/polymethyl methacrylate composite. **a** pouring hydroxyapatite slurry in a cold copper finger, **b** producing hydroxyapatite scaffold via freeze-drying and sintering methods, **c** compression of the obtained scaffold, **d** grafting of the scaffold by methacrylate groups, and **e** infiltration of methyl methacrylate solution through hydroxyapatite scaffold (Reprinted from [46] by permission from Wiley)

In an innovative procedure, Bai et al. [46] attempted to produce a nacre-mimetic composite from a hydroxyapatite composite. In this regard, as indicated in Fig. 2.6, they fabricated a hydroxyapatite scaffold via a bidirectional freeze-drying method. Hydroxyapatite slurry was poured on a cold copper finger covered by polydimethyl-siloxane wedges. The wedges created a temperature gradient perpendicular to the copper finger, and ice crystals started to grow in the directional of the gradient. Through the process hydroxyapatite particles entrapped in the free space of the grown ice crystals. The ice was sublimed via the freeze-drying method, and the remained structure was sintered at 1300 °C for 4 h. The obtained scaffold contained 70% porosity and 30% ceramic fraction. By applying unidirectional compaction, the porosity reduced to 15–25%, ceramic content increased to 75–85%, and hydroxyap-atite lamellar structure defragged into the separate bricks. The final scaffold structure was grafted with 3-(trimethoxysilyl)propyl methacrylate to unable the polymeriza-tion of polymethyl methacrylate on the densified scaffold. At the last step, methyl methacrylate monomer solution was infiltered through the scaffold, and the final hydroxyapatite/polymethyl methacrylate composite was achieved. The most impor-tant feature of this composite was, however, it contained more than 70 vol.% of ceramic part and it achieved distinguishable mechanical properties compared to the composite produced with other methods.

Radha et al. [47] combined the phase separation and wet chemical approaches to fabricate a polymethyl methacrylate scaffold containing hydroxyapatite. In this work, according to Fig. 2.7, polymethyl methacrylate was dissolved in acetone, and

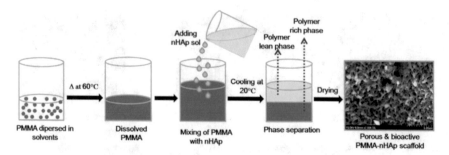

Fig. 2.7 Producing polymethyl methacrylate/nano-hydroxyapatite scaffold via a combined phase separation and wet chemical methods (Reprinted from [47] by permission from Elsevier)

then a solution of $Ca(NO_3)_2 \cdot 4H_2O$ and $(NH_4)_2HPO_4$ at pH of 10 was added to it. The mixture was stirred for 15 min and then left at 20 °C for 48 h until a porous scaffold produced.

2.4.3.2 Polypropylene Fumarate

Henslee et al. [48] produced a porous space maintainer practiced in craniofacial surgery. To this end, they added carboxymethylcellulose, as a porogen agent, to the mixture of polypropylene fumarate and n-vinyl pyrrolidine.

In another research, Kim et al. [49] fabricated an injectable scaffold via the gas foaming method. They added dry carbonate salt, $NaHCO_3$ or $CaCO_3$, as foaming agents to polypropylene fumarate and n-vinyl pyrrolidine solution. Citric acid and an initiator agent were added to the mixture. During stirring, the reaction between bicarbonate salt and weak acid generated CO_2 gases. The foamed polymer then injected by a syringe. Figure 2.8 shows micrographs of the obtained scaffold.

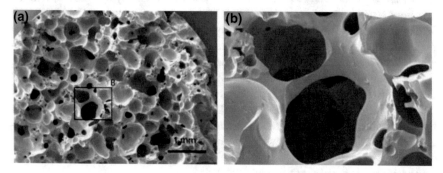

Fig. 2.8 Micrographs of polypropylene fumarate scaffold fabricated via gas foaming method (Reprinted from [49] by permission from Wiley)

Fisher et al. [50] used the photocross-linkable polypropylene fumarate to produce a scaffold. They used bis(2,4,6-trimethylbenzoyl) phenylphosphine oxide as a photoinitiator agent and exposed long wave ultra-violet light for 30 min to crosslink polypropylene fumarate. In this study, NaCl as a porogen agent was combined with the precursor compound before crosslinking step. The porogen agent was removed by immersing the obtained composite in water.

2.4.3.3 Calcium Phosphate Cement

Hesaraki et al. [51] fabricated a calcium phosphate cement scaffold via the gas foaming method. In this manner, the cement produced from a mixture of tetracalcium phosphate and dicalcium phosphate anhydrate powders with disodium hydrogen phosphate and sodium alginate solution. To produce the scaffold, sodium hydrogen carbonate, $NaHCO_3$, was mixed with the powder part and citric acid monohydrate, $C_6H_8O_7 \cdot H_2O$, added to the solution part. Chen et al. [52], by using calcium phosphate cement as produced by in Hesaraki's paper, fabricated a scaffold which contained mesenchymal stem cell-encapsulated microbeads.

In a kind of research, Vasconcellos and Alberto dos Santos [53] fabricated a calcium phosphate cement scaffold via the particulate leaching method. To obtain acceptable mechanical properties, the scaffold was reinforced by poly(lactic-co-glycolic) acid fibers. In this regard, α-tricalcium phosphate powders were mixed with poly(lactic-co-glycolic) acid fibers and paraffin spheres. The mixture was added to a solution containing $Na_2HPO_4 \cdot 12H_2O$. After the cement set, the cement was placed in an oven at 100 °C for 24 h to extract paraffin and produce the scaffold.

Lode et al. [54] produced a calcium phosphate cement scaffold via the 3-D printing method. The cement paste was obtained from mixing of a powder mixture containing α-tricalcium phosphate, dicalcium phosphate anhydride, calcium carbonate, and hydroxyapatite with a carrier liquid including short-chain triglyceride, Polysorbate 80, and phosphoric acid monohexadecyl ester. The scaffold was produced via deposition of the paste from the 3-D printing nozzle and then transferred to water at 37 °C for three days. The printing device is indicated in Fig. 2.9, and the final obtained scaffold from a different view is shown in Fig. 2.10. Because the process was done at room temperature, Akkineni et al. [55] can load bioactive agents including Bovine serum albumin and vascular endothelial growth factor encapsulated in chitosan/dextran sulfate microparticles to the scaffold through the printing process.

The particulate leaching technique was used by Liu et al. [56] to produce a calcium phosphate cement scaffold carrying mesenchymal stem cells. In this regard, mannitol crystal, that is soluble in water, was used as the porogen agent. To protect cells from the harsh environment during cementation reaction, they were encapsulated in chitosan/β-glycerophosphate hydrogel and added to the cement during the scaffold fabrication.

Bian et al. [57] produced a scaffold that mimicked the real structure of bone consisting of a tubular osteon structure. In this work, they produced osteon cylinders from a membrane of small intestinal submucosa coated by the calcium phosphate

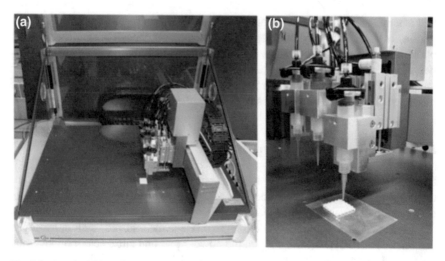

Fig. 2.9 A typical 3-D printing device that deposits a paste from a syringe on a platform (Reprinted from [54] by permission from John Wiley and Sons)

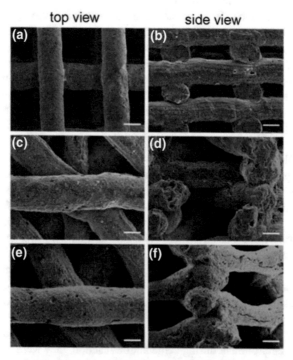

Fig. 2.10 Micrographs of a calcium phosphate cement scaffold produced via the 3-D printing method (Reprinted from [54] by permission from John Wiley and Sons)

cement. Small intestinal submucosa is a collagenous extracellular matrix containing various cytokines, which induces vascularization and cell growth. By rolling the coated membrane, an osteon-mimicked structure was produced layer-by-layer (as shown in Fig. 2.11).

In addition to scaffolds built in a laboratory environment and then transferred to the patient body, some of them can be injected directly. For instance, Montufar et al. [58] fabricated a calcium phosphate cement scaffold by using the foamed surfactant solution. In this method as shown in Fig. 2.12, first, polysorbate 80 was added to water and by applying agitation, a foamed solution was obtained. Then α-tricalcium phosphate powders were mixed with the foamed solution, and the mixture was injected to a defect site. The cementation reaction gradually occurred, and a porous structure remained. Moreover, in another research [59], the same group applied gelatin

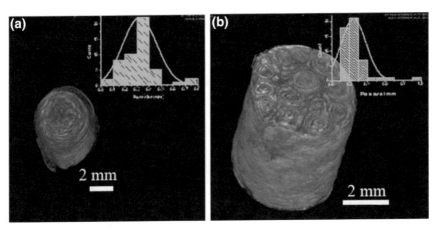

Fig. 2.11 Micro-computed tomography images of **a** osteon-mimicked and **b** bone-mimicked scaffold (Reprinted from [57] by permission from Elsevier)

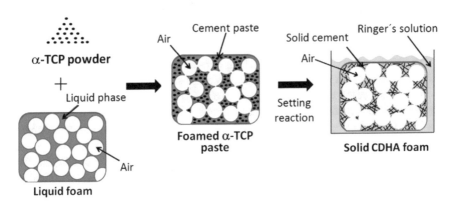

Fig. 2.12 Schematic of producing a calcium phosphate cement scaffold via using a foamed surfactant solution (Reprinted from [58] by permission from Elsevier)

Fig. 2.13 Micrographs of a
calcium phosphate cement
scaffold produced with a
gelatin-foamed solution
(Reprinted from [59] by
permission from Springer
Nature)

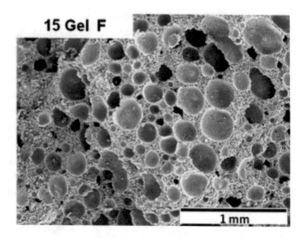

as a foaming agent and produced a calcium phosphate cement scaffold, as shown in
Fig. 2.13.

Alge et al. [60] reinforced calcium phosphate cement by polypropylene fumarate
cement. In this work, they fabricated a scaffold from calcium phosphate cement via
the 3-D printing method and immersed it in the solution of polypropylene fumarate
cement. The obtained final compressive strength was seven times higher than the
plain scaffold.

2.4.3.4 Magnesium Phosphate Cement

Meininger et al. [61, 62] evaluated the fabrication of magnesium phosphate cement
scaffold via the 3-D powder printing method. In their work, strontium was doped in
precursor powders of magnesium phosphate to achieve the aim of improving compressive strength and radiopacity of the cement. Furthermore, adding strontium can
show a side effect in decreasing osteoclastogenesis and increasing osteoblast proliferation. In this manner, for each layer 100 μm of strontium-doped magnesium phosphate powder covered the stage, and ammonium dihydrogen phosphate was used as its
binder. As shown in Fig. 2.14 obtained by micro-computed tomography, the scaffolds
containing strontium possess more interconnected structure than the plain one.

In another research, a Korean group [63, 64] produced struvite scaffold via
the 3-D printing method and paste extrusion deposition. The group fabricated the
scaffold in a two-step process (as shown in Fig. 2.15) including producing green
body scaffold and cementation. Through the first step, $Mg_3(PO_4)_2$ powders were
mixed by 1% hydroxypropyl methyl cellulose and dissolved in 30% ethanol. The
green scaffold was deposited from a syringe nozzle based on a shape designed in
the computer. After drying the green body at ambient temperature, in the second
step, the green body was immersed in diammonium hydrogen phosphate solution
to induce the cementation process. After a day, the obtained body was washed with

$Mg_3(PO_4)_2$ $Mg_{2.5}Sr_{0.5}(PO_4)_2$ $Mg_2Sr_1(PO_4)_2$

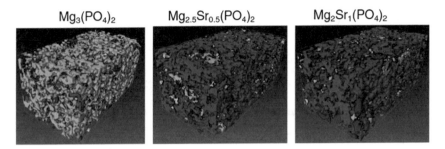

Fig. 2.14 Pore structure of magnesium phosphate cement scaffold obtained via micro-computed tomography. The blue color shows more interconnected pore structure than green one (Reprinted from [62] by permission from Elsevier)

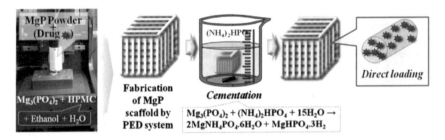

Fig. 2.15 3-D printing of magnesium phosphate cement scaffold (Reprinted from [63] by permission from Elsevier)

water and dried in ambient temperature. Because the process was done at room temperature, it is possible to add bioactive agents in the fabrication process. The group added lysozyme as a drug model and KR-34893 to induce the expression of osteoblast differentiation marker genes.

To more mimicking bone structure, adding bioactive agents and cells should be considered which is discussed in the next two chapters.

References

1. T. Nakamura, M. Takemoto, Osteoconduction and its evaluation, in *Bioceramics and Their Clinical Applications*, ed. by T. Kokubo (CRC Press, Cambridge, 2008), pp. 183–198
2. M.M. Stevens, Biomaterials for bone tissue engineering. Mater. Today **11**, 18–25 (2008). https://doi.org/10.1016/S1369-7021(08)70086-5
3. U.G.K. Wegst, H. Bai, E. Saiz, A.P. Tomsia, R.O. Ritchie, Bioinspired structural materials. Nat. Mater. **14**, 23–36 (2015). https://doi.org/10.1038/nmat4089
4. A. Shekaran, A.J. García, Extracellular matrix-mimetic adhesive biomaterials for bone repair. J. Biomed. Mater. Res.—Part A **96**(1), 261–272 (2011). https://doi.org/10.1002/jbm.a.32979
5. R.I. Freshney, *Culture of Animal Cells: A Manual of Basic Technique and Specialized Applications*, 6th edn. (Wiley, Hoboken, 2010)

6. G. Karp, *Cell and Molecular Biology: Concepts and Experiments*, 7th edn. (Wiley, Danvers, 2013)
7. R. Vaishya, M. Chauhan, A. Vaish, Bone cement. J. Clin. Orthop. Trauma **4**, 157–163 (2013). https://doi.org/10.1016/j.jcot.2013.11.005
8. C. Duval-Terrié, L. Lebrun, Polymerization and characterization of PMMA. Polymer chemistry laboratory experiments for undergraduate students. J. Chem. Educ. **83**, 443 (2006). https://doi.org/10.1021/ed083p443
9. J. Hasenwinkel, Bone cement, in *Encyclopedia of Biomaterials and Biomedical Engineering*, 2nd edn., ed. by G.E. Wnek, G.L. Bowlin (Informa Healthcare, New YorK, 2008), pp. 403–412
10. G. Lewis, Properties of acrylic bone cement: state of the art review. J. Biomed. Mater. Res. **38**, 155–182 (1997). https://doi.org/10.1002/(SICI)1097-4636(199722)38:2%3c155:AID-JBM10%3e3.0.CO;2-C
11. L. Hernández, M. Gurruchaga, I. Goñi, Injectable acrylic bone cements for vertebroplasty based on a radiopaque hydroxyapatite. Formulation and rheological behaviour. J. Mater. Sci. Mater. Med. **20**, 89–97 (2009). https://doi.org/10.1007/s10856-008-3542-y
12. Y. Wang, Y. Xiao, X. Huang, M. Lang, Preparation of poly(methyl methacrylate) grafted hydroxyapatite nanoparticles via reverse ATRP. J. Colloid Interface Sci. **360**, 415–421 (2011). https://doi.org/10.1016/j.jcis.2011.04.093
13. L. Chen, D. Zhai, Z. Huan, N. Ma, H. Zhu, C. Wu, J. Chang, Silicate bioceramic/PMMA composite bone cement with distinctive physicochemical and bioactive properties. RSC Adv. **5**, 37314–37322 (2015). https://doi.org/10.1039/C5RA04646G
14. B. Marrs, R. Andrews, T. Rantell, D. Pienkowski, Augmentation of acrylic bone cement with multiwall carbon nanotubes. J. Biomed. Mater. Res., Part A **77A**, 269–276 (2006). https://doi.org/10.1002/jbm.a.30651
15. K. Król, K. Pielichowska, Modification of acrylic bone cements by poly(ethylene glycol) with different molecular weight. Polym. Adv. Technol. **27**, 1284–1293 (2016). https://doi.org/10.1002/pat.3792
16. Z. He, Q. Zhai, M. Hu, C. Cao, J. Wang, H. Yang, B. Li, Bone cements for percutaneous vertebroplasty and balloon kyphoplasty: current status and future developments. J. Orthop. Transl. **3**, 1–11 (2015). https://doi.org/10.1016/j.jot.2014.11.002
17. M. Salarian, W.Z. Xu, M.C. Biesinger, P.A. Charpentier, Synthesis and characterization of novel TiO 2 -poly(propylene fumarate) nanocomposites for bone cementation. J. Mater. Chem. B **2**, 5145–5156 (2014). https://doi.org/10.1039/C4TB00715H
18. E.L.S. Fong, B.M. Watson, F.K. Kasper, A.G. Mikos, Building bridges: leveraging interdisciplinary collaborations in the development of biomaterials to meet clinical needs. Adv. Mater. **24**, 4995–5013 (2012). https://doi.org/10.1002/adma.201201762
19. S.J. Peter, P. Kim, A.W. Yasko, M.J. Yaszemski, A.G. Mikos, Crosslinking characteristics of an injectable poly(propylene fumarate)/beta-tricalcium phosphate paste and mechanical properties of the crosslinked composite for use as a biodegradable bone cement. J. Biomed. Mater. Res. **44**, 314–321 (1999)
20. N.S. Anitha, V. Thomas, M. Jayabalan, Poly(propylene fumarate)ln-vinyl pyrrolidone copolymer-based bone cement: setting and in-vitro biodegradation. J. Indian Inst. Sci. **79**, 431–442 (1999)
21. S. He, M.J. Yaszemski, A.W. Yasko, P.S. Engel, A.G. Mikos, Injectable biodegradable polymer composites based on poly(propylene fumarate) crosslinked with poly(ethylene glycol)-dimethacrylate. Biomaterials **21**, 2389–2394 (2000). https://doi.org/10.1016/S0142-9612(00)00106-X
22. J.P. Fisher, D. Dean, A.G. Mikos, Photocrosslinking characteristics and mechanical properties of diethyl fumarate/poly(propylene fumarate) biomaterials. Biomaterials **23**, 4333–4343 (2002)
23. D. Hakimimehr, D.-M. Liu, T. Troczynski, In-situ preparation of poly(propylene fumarate)–hydroxyapatite composite. Biomaterials **26**, 7297–7303 (2005). https://doi.org/10.1016/j.biomaterials.2005.05.065
24. X. Shi, B. Sitharaman, Q.P. Pham, F. Liang, K. Wu, W.E. Billups, L.J. Wilson, A.G. Mikos, Fabrication of porous ultra-short single-walled carbon nanotube nanocomposite scaffolds for bone

tissue engineering. Biomaterials **28**, 4078–4090 (2007). https://doi.org/10.1016/j.biomaterials. 2007.05.033

25. B. Sitharaman, X. Shi, X.F. Walboomers, H. Liao, V. Cuijpers, L.J. Wilson, A.G. Mikos, J.A. Jansen, In vivo biocompatibility of ultra-short single-walled carbon nanotube/biodegradable polymer nanocomposites for bone tissue engineering. Bone **43**, 362–370 (2008). https://doi. org/10.1016/j.bone.2008.04.013

26. E. Fernandez, F.J. Gil, M.P. Ginebra, F.C.M. Driessens, J.A. Planell, S.M. Best, Calcium phosphate bone cements for clinical applications. Part II: Precipitate formation during setting reactions. J. Mater. Sci. Mater. Med. **10**, 177–183 (1999). https://doi.org/10.1023/A: 1008989525461

27. M. Ginebra, E. Fernandez, F.C.M. Driessens, J.A. Planell, Modeling of the hydrolysis of a-tricalcium phosphate. J. Am. Ceram. Soc. **82**, 2808–2812 (1999)

28. A. Ewald, K. Helmschrott, G. Knebl, N. Mehrban, L.M. Grover, U. Gbureck, Effect of cold-setting calcium- and magnesium phosphate matrices on protein expression in osteoblastic cells. J. Biomed. Mater. Res. B Appl. Biomater. **96**, 326–332 (2011). https://doi.org/10.1002/jbm.b. 31771

29. C. Großardt, A. Ewald, L.M. Grover, J.E. Barralet, U. Gbureck, Passive and active in vitro resorption of calcium and magnesium phosphate cements by osteoclastic cells. Tissue Eng. Part A **16**, 3687–3695 (2010). https://doi.org/10.1089/ten.tea.2010.0281

30. U. Klammert, A. Ignatius, U. Wolfram, T. Reuther, U. Gbureck, In vivo degradation of low temperature calcium and magnesium phosphate ceramics in a heterotopic model. Acta Biomater. **7**, 3469–3475 (2011). https://doi.org/10.1016/j.actbio.2011.05.022

31. F. Tamimi, Z. Sheikh, J. Barralet, Dicalcium phosphate cements: brushite and monetite. Acta Biomater. **8**, 474–487 (2012). https://doi.org/10.1016/j.actbio.2011.08.005

32. K.L. Low, S.H. Tan, S.H.S. Zein, J.A. Roether, V. Mouriño, A.R. Boccaccini, Calcium phosphate-based composites as injectable bone substitute materials. J. Biomed. Mater. Res. B Appl. Biomater. **94**, 273–286 (2010). https://doi.org/10.1002/jbm.b.31619

33. D.L. Alge, W.S. Goebel, T.-M.G. Chu, Effects of DCPD cement chemistry on degradation properties and cytocompatibility: comparison of MCPM/β-TCP and MCPM/HA formulations. Biomed. Mater. **8**, 025010 (2013). https://doi.org/10.1088/1748-6041/8/2/025010

34. F. Chen, C. Liu, J. Wei, X. Chen, Physicochemical properties and biocompatibility of white dextrin modified injectable calcium-magnesium phosphate cement. Int. J. Appl. Ceram. Technol. **9**, 979–990 (2012). https://doi.org/10.1111/j.1744-7402.2011.02705.x

35. G. Mestres, M.-P. Ginebra, Novel magnesium phosphate cements with high early strength and antibacterial properties. Acta Biomater. **7**, 1853–1861 (2011). https://doi.org/10.1016/j.actbio. 2010.12.008

36. C. Moseke, V. Saratsis, U. Gbureck, Injectability and mechanical properties of magnesium phosphate cements. J. Mater. Sci. Mater. Med. **22**, 2591–2598 (2011). https://doi.org/10.1007/ s10856-011-4442-0

37. M. Nabiyouni, T. Brückner, H. Zhou, U. Gbureck, S.B. Bhaduri, Magnesium-based bioceramics in orthopedic applications. Acta Biomater. **66**, 23–43 (2017). https://doi.org/10.1016/j.actbio. 2017.11.033

38. N. Ostrowski, A. Roy, P.N. Kumta, Magnesium phosphate cement systems for hard tissue applications: a review. ACS Biomater. Sci. Eng. **2**, 1067–1083 (2016). https://doi.org/10.1021/ acsbiomaterials.6b00056

39. A. Ewald, K. Helmschrott, G. Knebl, N. Mehrban, L.M. Grover, U. Gbureck, Effect of cold-setting calcium- and magnesium phosphate matrices on protein expression in osteoblastic cells. J. Biomed. Mater. Res.—Part B Appl. Biomater. **96B**, 326–332 (2011). https://doi.org/10.1002/ jbm.b.31771

40. U. Klammert, T. Reuther, M. Blank, I. Reske, J.E. Barralet, L.M. Grover, A.C. Kübler, U. Gbureck, Phase composition, mechanical performance and in vitro biocompatibility of hydraulic setting calcium magnesium phosphate cement. Acta Biomater. **6**, 1529–1535 (2010). https:// doi.org/10.1016/j.actbio.2009.10.021

41. S.M. Kenny, M. Buggy, Bone cements and fillers: a review. J. Mater. Sci. Mater. Med. **14**, 923–938 (2003). https://doi.org/10.1023/A:1026394530192

42. Q.-Z. Chen, A.R. Boccaccini, Bioactive materials and scaffolds for tissue engineering, in *Encyclopedia of Biomaterials and Biomedical Engineering*, 2nd edn., ed. by G.E. Wnek, G.I. Bowlin (Informa Healthcare, New YorK, 2008), pp. 142–151

43. E.D. Boland, P.G. Espy, G.L. Bowlin, Tissue engineering scaffolds, in *Encyclopedia of Biomaterials and Biomedical Engineering*, 2nd edn., ed. by G.E. Wnek, G.I. Bowlin (Informa Healthcare, New YorK, 2008), pp. 2828–2837

44. T. Garg, O. Singh, S. Arora, R. Murthy, Scaffold: a novel carrier for cell and drug delivery. Crit. Rev. Ther. Drug Carrier Syst. **29**, 1–63 (2012)

45. M. Shi, J.D. Kretlow, P.P. Spicer, Y. Tabata, N. Demian, M.E. Wong, F.K. Kasper, A.G. Mikos, Antibiotic-releasing porous polymethylmethacrylate/gelatin/antibiotic constructs for craniofacial tissue engineering. J. Control Release **152**, 196–205 (2011). https://doi.org/10.1016/j.jconrel.2011.01.029

46. H. Bai, F. Walsh, B. Gludovatz, B. Delattre, C. Huang, Y. Chen, A.P. Tomsia, R.O. Ritchie, Bioinspired hydroxyapatite/poly(methyl methacrylate) composite with a nacre-mimetic architecture by a bidirectional freezing method. Adv. Mater. **28**, 50–56 (2016). https://doi.org/10.1002/adma.201504313

47. G. Radha, S. Balakumar, B. Venkatesan, E. Vellaichamy, A novel nano-hydroxyapatite— PMMA hybrid scaffolds adopted by conjugated thermal induced phase separation (TIPS) and wet-chemical approach: analysis of its mechanical and biological properties. Mater. Sci. Eng. C **75**, 221–228 (2017). https://doi.org/10.1016/j.msec.2016.12.133

48. A.M. Henslee, S.R. Shah, M.E. Wong, A.G. Mikos, F.K. Kasper, Degradable, antibiotic releasing poly(propylene fumarate)-based constructs for craniofacial space maintenance applications. J. Biomed. Mater. Res., Part A **103**, 1485–1497 (2015). https://doi.org/10.1002/jbm.a.35288

49. C.W. Kim, R. Talac, L. Lu, M.J. Moore, B.L. Currier, M.J. Yaszemski, Characterization of porous injectable poly-(propylene fumarate)-based bone graft substitute. J. Biomed. Mater. Res., Part A **85A**, 1114–1119 (2008). https://doi.org/10.1002/jbm.a.31633

50. J.P. Fisher, T.A. Holland, D. Dean, P.S. Engel, A.G. Mikos, Synthesis and properties of photocross-linked poly(propylene fumarate) scaffolds. J. Biomater. Sci. Polym. Ed. **12**, 673–687 (2001). https://doi.org/10.1163/156856201316883476

51. S. Hesaraki, F. Moztarzadeh, D. Sharifi, Formation of interconnected macropores in apatitic calcium phosphate bone cement with the use of an effervescent additive. J. Biomed. Mater. Res., Part A **83A**, 80–87 (2007). https://doi.org/10.1002/jbm.a.31196

52. W. Chen, H. Zhou, M. Tang, M.D. Weir, C. Bao, H.H.K. Xu, Gas-foaming calcium phosphate cement scaffold encapsulating human umbilical cord stem cells. Tissue Eng. Part A **18**, 816–827 (2012). https://doi.org/10.1089/ten.tea.2011.0267

53. L.A. Vasconcellos, L.A. dos Santos, Calcium phosphate cement scaffolds with PLGA fibers. Mater. Sci. Eng., C **33**, 1032–1040 (2013). https://doi.org/10.1016/j.msec.2012.10.019

54. A. Lode, K. Meissner, Y. Luo, F. Sonntag, S. Glorius, B. Nies, C. Vater, F. Despang, T. Hanke, M. Gelinsky, Fabrication of porous scaffolds by three-dimensional plotting of a pasty calcium phosphate bone cement under mild conditions. J. Tissue Eng. Regen. Med. **8**, 682–693 (2014). https://doi.org/10.1002/term.1563

55. A.R. Akkineni, Y. Luo, M. Schumacher, B. Nies, A. Lode, M. Gelinsky, 3D plotting of growth factor loaded calcium phosphate cement scaffolds. Acta Biomater. **27**, 264–274 (2015). https://doi.org/10.1016/j.actbio.2015.08.036

56. T. Liu, J. Li, Z. Shao, K. Ma, Z. Zhang, B. Wang, Y. Zhang, Encapsulation of mesenchymal stem cells in chitosan/β-glycerophosphate hydrogel for seeding on a novel calcium phosphate cement scaffold. Med. Eng. Phys. **56**, 9–15 (2018). https://doi.org/10.1016/j.medengphy.2018.03.003

57. T. Bian, K. Zhao, Q. Meng, H. Jiao, Y. Tang, J. Luo, Fabrication and performance of calcium phosphate cement/small intestinal submucosa composite bionic bone scaffolds with different microstructures. Ceram. Int. **44**, 9181–9187 (2018). https://doi.org/10.1016/j.ceramint.2018.02.127

58. E.B. Montufar, T. Traykova, C. Gil, I. Harr, A. Almirall, A. Aguirre, E. Engel, J.A. Planell, M.P. Ginebra, Foamed surfactant solution as a template for self-setting injectable hydroxyapatite scaffolds for bone regeneration. Acta Biomater. **6**, 876–885 (2010). https://doi.org/10.1016/j. actbio.2009.10.018

59. E.B. Montufar, T. Traykova, E. Schacht, L. Ambrosio, M. Santin, J.A. Planell, M.-P. Ginebra, Self-hardening calcium deficient hydroxyapatite/gelatine foams for bone regeneration. J. Mater. Sci. Mater. Med. **21**, 863–869 (2010). https://doi.org/10.1007/s10856-009-3918-7

60. D.L. Alge, J. Bennett, T. Treasure, S. Voytik-Harbin, W.S. Goebel, T.-M.G. Chu, Poly(propylene fumarate) reinforced dicalcium phosphate dihydrate cement composites for bone tissue engineering. J. Biomed. Mater. Res. A **100**, 1792–1802 (2012). https://doi.org/10. 1002/jbm.a.34130

61. S. Meininger, C. Moseke, K. Spatz, E. März, C. Blum, A. Ewald, E. Vorndran, Effect of strontium substitution on the material properties and osteogenic potential of 3D powder printed magnesium phosphate scaffolds. Mater. Sci. Eng., C **98**, 1145–1158 (2019). https://doi.org/10. 1016/j.msec.2019.01.053

62. S. Meininger, S. Mandal, A. Kumar, J. Groll, B. Basu, U. Gbureck, Strength reliability and in vitro degradation of three-dimensional powder printed strontium-substituted magnesium phosphate scaffolds. Acta Biomater. **31**, 401–411 (2016). https://doi.org/10.1016/j.actbio.2015. 11.050

63. J. Lee, M.M. Farag, E.K. Park, J. Lim, H. Yun, A simultaneous process of 3D magnesium phosphate scaffold fabrication and bioactive substance loading for hard tissue regeneration. Mater. Sci. Eng., C **36**, 252–260 (2014). https://doi.org/10.1016/j.msec.2013.12.007

64. J.A. Kim, H. Yun, Y.-A. Choi, J.-E. Kim, S.-Y. Choi, T.-G. Kwon, Y.K. Kim, T.-Y. Kwon, M.A. Bae, N.J. Kim, Y.C. Bae, H.-I. Shin, E.K. Park, Magnesium phosphate ceramics incorporating a novel indene compound promote osteoblast differentiation in vitro and bone regeneration in vivo. Biomaterials **157**, 51–61 (2018). https://doi.org/10.1016/j.biomaterials.2017.11.032

Chapter 3
Productivity: Cells

3.1 Introduction

Bone tissue engineering is an approach that mimics the natural tissue condition; so, it needs three essential components. Cells are the first component that enables regenerativity of the approach. The second one is growth and differentiation factors that favor appropriate cells to grow in the right position. The last but not the least is the scaffold that plays its role as a substrate for cell activities, including attachment, proliferation, and differentiation [1].

Although there is a controversial discussion about the selection of the most effective cell types, all scientists agree that cell usage is influential in bone regenerative methods. In the case of bone tissue engineering, the selection is established based on what happens during bone repair, naturally. The event involves three stages: (1) expression of essential agents in vascularization and bone regeneration, (2) establishing a new substrate for employing attributed cells, and (3) creating a bone structure with the embedded required blood vessels [2]. The impressive attempts of both surgeons and scientists to apply a multi-component strategy in curing large bone defects indicate the importance of the mentioned stages. Therefore, in this chapter, the so-called bone tissue engineering is considered (Fig. 3.1) [3].

Using scaffolds, naked or accompanied by growth factors, are also applied to induce presenting cells in a particular situation to regenerate bone, but this strategy is not effective in large bone defects. To overcome this shortage, employing cells attract much attention. The method not only favors in the formation of new bone tissue but also enables the bone to tolerate what would happen in the future [4].

3.2 Bone Cells

In this section, cells embedded in the bone structure are introduced.

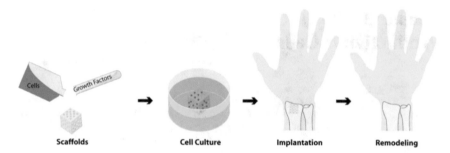

Fig. 3.1 Schematic illustration of a multi-component strategy for curing bone defects. The strategy involves using cells, factors, and scaffolds (Reprinted from [3] by permission from Elsevier)

3.2.1 Stem Cell Niches in Bone

Generally, the bone tissue is constructed from different components, including bone cells (such as pre-osteoblasts, osteoblasts, and osteocytes), collagen fibrils, and some mineral compounds which consist of calcium and phosphate. Besides that bone marrow and vessels fill the bone cavity, they are the place of inhabiting of two critical stem cell types, including hematopoietic stem cells and mesenchymal stem cells (Fig. 3.2). Hematopoietic stem cells place among the stromal cells, and their roles are to produce immune, blood, and osteoclast cells. The other stem cell type, mesenchymal stem cells, is the source of osteoblasts, adipocytes, chondrocytes, fibroblast, and other stromal cells. These cells are classified as mesenchymal lineage cells. The activity of both stem cell types is essential in bone homeostasis and cellular generation [5, 6].

3.2.2 Resident Cells in the Bone

Four kinds of cells are resident in the bone such as osteoblasts, osteoclasts, osteocytes, and osteoprogenitor cells. Each of them has individual roles in bone life. Bone regeneration is the responsibility of osteoblast cells. The cells differentiate from daughter cells originated from osteoprogenitor cells. On the other hand, bone structure can be dissolved by lysosomal enzymes and acids released by giant bone cells, called the osteoclast cells. The cells can contain even up to 50 nuclei. Most bone matrices are occupied by the mature bone cells named osteocytes (see Fig. 3.3).

3.2.2.1 Osteoblasts

Osteoblasts, recognized as their bone forming function, possess some characteristics: (1) cuboidal morphology; (2) covering bone surface; (3) making 4–6% of the total embedded cells in bone; (4) knowing as protein synthesizing cells due to their

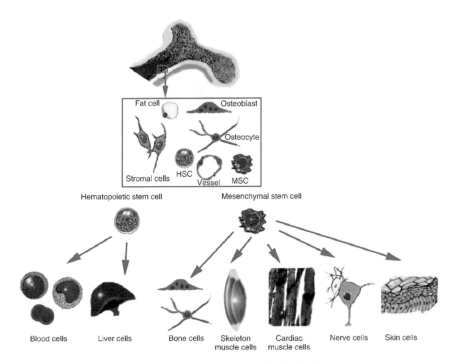

Fig. 3.2 The schematic of natural bone and its embedded cells; besides other organ cells that can differentiate from stem cells inhabiting in bone (Reprinted from [7] by permission from Elsevier)

rough endoplasmic reticulum, prominent Golgi apparatus, and different secretory vesicles; (5) secretor of osteoid in bone; and (6) differentiated from mesenchymal stem cells [8]. The bone formation process by the cells consists of two main steps including secretion of organic matrix and mineralization. In the former step, collagen proteins, mainly type I collagen, non-collagen proteins (OCN, osteonectin, BSP II and osteopontin) and proteoglycans such as decorin and biglycan are secreted. In the mineralization step, within the matrix vesicles, originated from the external membrane of osteoblasts, odontoblasts, and chondrocytes, hydroxyapatite crystals are formed. Unfortunately, scientists have not yet achieved to detailed information about this step [9].

3.2.2.2 Osteocytes

A fully matured type of cells in the bone structure, differentiated from osteoblasts, is osteocytes. From its characteristics, it can be mentioned that the cells make up 90–95% of the total bone cells, lives for decades, and their configuration is in the dendritic form [6]. The presence of the cells is critical in bone metabolism. The cells are the resource of osteocalcin, galectin 3, and CD44, a cell adhesion receptor for

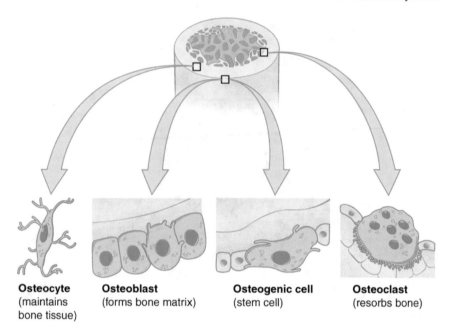

Osteocyte   **Osteoblast**   **Osteogenic cell**   **Osteoclast**
(maintains   (forms bone matrix)   (stem cell)   (resorbs bone)
bone tissue)

Fig. 3.3 The schematic and placement of four main cell types, including osteoprogenitor cell, osteoblast, osteoclast, and osteocyte, in bone tissue (Reprinted from [10], CC BY 4.0)

hyaluronate. In addition, some important proteins that cause adhesion of cells and control the mineral concentration of bone fluid within lacunae and the canalicular network are expressed by osteocytes. The connection of the cells and between the cells and bone surface maintains by multiple filipodia placed on the surface of the cells. Integral cellular proteins, connexins, fill the intercellular space, and control direct communication of the cells. The proteins connect the cells, both metabolically and electrically, and play their vital role in maturation, activity, and survival of osteocytes. Despite the mentioned roles of osteocytes, the presence of lysosomes in the cells gives them phagocytotic function that gets activated during the osteolysis process [11].

3.2.2.3 Osteoclasts

Another kind of the bone cells, osteoclasts, differentiates from the hematopoietic lineage, which is the precursor of macrophage cells. The most important role of the cells is bone resorbing during bone remodeling. Two factors including hydrogen ions and cathepsin K, secreted by osteoclasts, are present in the remodeling. The former decomposes the mineral bone matrix by acidifying the resorption compartment and the latter digests insoluble parts of the matrix such as collagen. The integrin placed on the surface of the cells creates an appropriate situation for bonding with the bone matrix. During the bonding, osteoclasts get polarized and create a defined space between the cells and the matrix. The mentioned factors would be released in this defined space [6, 12].

3.2.2.4   Osteoprogenitor Cells

A small fraction of the mesenchymal cells consists of osteoprogenitor cells that are the source of osteoblasts and preserve their population in the bone remodeling process. The cells located in three main sites, including the periosteum and endosteum of bone in addition to lining of passageways such as vessels [13].

3.3   Cell Sources for Bone Tissue Engineering

The feasibility of the implanting cells in bone defects and their ability to produce osteoblasts and neo-vasculature are the main parameters that determine the efficiency of cellular therapy [2]. In Table 3.1, some benefits and drawbacks of different stem cell sources are listed. In the following, the ability of the sources for bone treatment is evaluated.

3.3.1   Embryonic Stem Cells

Human embryonic stem cells are extracted from the inner cell mass of a blastocyst stage embryo, as shown in Fig. 3.4, via the microsurgical procedure. The outstanding feature of the cells is their strong capability for proliferation and differentiation. They can proliferate for a long time without differentiation and also they can differentiate to all three embryonic germ layers, called ectoderm, mesoderm, and endoderm. Because of these features, the cells are classified as pluripotent cells [14, 15]. Osteoblasts can be differentiated from these cells by adding some ingredients consisting of ascorbic acid, β-glycerophosphate, and dexamethasone. Due to the risk of carcinogenicity the clinical applications of human embryonic stem cells are restricted [16].

By seeding human stem cells in calcium phosphate cement scaffold, the expression of osteogenic markers, such as alkaline phosphatase, osteocalcin, collagen I, and Runx2, was reported. The event confirmed the osteogenesis ability of the scaffold [18]. By implanting the scaffold in cranial defects in rats, bone growth and vascularization improved compared with the plain scaffold [19]. In some researches, hydrogel fibers encapsulating human embryonic stem cells were embedded in calcium phosphate cement. The encapsulation preserved the cells from the harsh environment of the cement paste in addition to maintained injectability of the paste. In this situation, the scaffold also indicated an acceptable release of osteogenic markers [14, 20].

Table 3.1 Benefits and drawbacks of different stem cell sources for bone tissue engineering (Reprinted from [17], CC BY 4.0)

Cell sources	Advantages	Disadvantages
Bone marrow-derived mesenchymal stem cells (BM-MSCs)	(i) High osteogenic potential (ii) Studied extensively	Low abundance; requires extensive in vitro expansion
Adipose-derived stem cells (ASCs)	(i) Similar osteogenic characteristics as BM-MCSs (ii) Highly abundant; easy to harvest surgically	More studies are needed to test their use in bone repair
Embryonic stem cells (ESCs)	(i) Pluripotency (ii) Capable of differentiating into all cell types in bone	(i) Ethical and regulatory constraints (ii) Produce teratomas when transplanted in vivo
Umbilical cord blood mesenchymal stem cells (CB-MSCs)	(i) High availability (ii) Broad differentiation and proliferation potential (iii) Higher in vivo safety than embryonic stem cells	(i) More difficult to be isolated than MSCs from the marrow (ii) More studies are needed to test their use in bone repair
Induced pluripotent stem cells (iPSCs)	(i) Pluripotency (ii) Capable of differentiating into all cell types in bone	(i) Reprogramming efficiency is low (ii) Require extensive expansion (iii) Safety concerns; limited clinical application
Adipose-derived stromal vascular fraction (SVF)	(i) Abundant; easily harvested via liposuction (ii) Able to form vascularized bone	(i) Cell population varies among donors (ii) 2–3-hour multistep isolation process

3.3.2 Induced Pluripotent Stem Cells

A new kind of stem cells without any ethical issues is the induced pluripotent stem cell. The cells are obtained by applying the transduction process and expression of transcription factors, such as Oct4, Sox2, cMyc, and Klf4, on somatic cells (Fig. 3.5) [21]. Although by using autologous cells, the possibility of immune rejection is reduced, the carcinogenicity of induced pluripotent stem cells exerts influence on the clinical application of the cells [16].

The seeding human induced pluripotent stem cells originated from mesenchymal stem cells on calcium phosphate cement scaffold indicated a great improvement in the induction of the bone and vascular growth [23, 24]. In addition, using biofunctionalized cement instead of the mentioned plain cement enhanced much more bone regeneration [25]. In another research, Wang et al. [26] applied the cells on macroporous calcium phosphate cement and implanted the graft in rat cranial defect. They reported a significant increase in the bone growth compared to cell-free cement (as shown in Fig. 3.6).

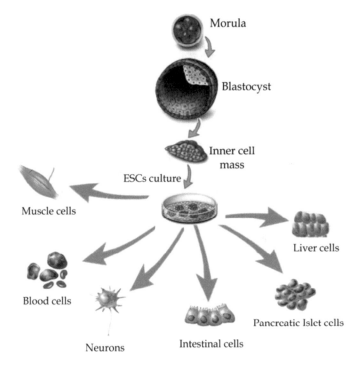

Fig. 3.4 Different cell types that can be differentiated from embryonic stem cells (Reprinted from [15], CC BY-NC-SA 3.0)

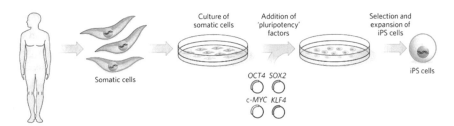

Fig. 3.5 The schematic of transduction of fibroblasts (a type of adult somatic cell) with pluripotency factors (SOX2, KLF4, c-MYC, and OCT4) to obtain induced pluripotent stem cells (Adapted from [22] by permission from Springer Nature)

Even the architectural aspects of a scaffold can affect cell activities. Sladkova et al. [27, 28] determined the optimum pore size of calcium phosphate cement scaffold to induce differentiation of human induced pluripotent stem cells to osteoblastic phenotype. In another study, the synergic effect of seeding endothelial cells and human induced pluripotent stem cells on calcium phosphate cement scaffold was assessed. The in vivo evaluation of the scaffolds after 12 weeks indicated the growth of 46% more bone tissue in the co-cultured scaffold compared to mono-cultured one.

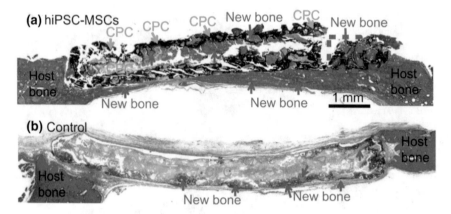

Fig. 3.6 The histological images of **a** co-cultured and **b** plain calcium phosphate cement scaffolds captured after implanting for 12 weeks. The red area is bone tissue grown in the scaffold and pointed by blue arrows (Adapted from [26] by permission from Elsevier)

The variation originated from induction of vascularization caused by endothelial cells besides bone growth induction of the stem cells. In general, the bone growth on the cell-seeded scaffold was significantly higher than the plain scaffold [29].

3.3.3 Fetal Stem Cells

Another kind of stem cell is fetal stem cells that compared to adult cells show outstanding features such as high withstanding at low oxygen condition, proliferation in a shorter time, and a higher ability in phenotypic potential. Also, the expression of a high level of angiogenic and trophic factors by fetal stem cells enhances the regeneration of the tissue surrounded the implanting site. The most attractive sources of fetal stem cells are the placenta, amniotic fluid, and umbilical cord blood, as shown in Fig. 3.7. The critical feature of these sources is the accessibility to the cells via a minimally invasive manner from the fetus. Umbilical cord blood can be detached and stored in cord blood banks after childbirth [30].

Among all sources, the amniotic fluid becomes an individual source of various cells because of contacting the fluid with the most part of a fetus during gestation. The available cells in the fluid are as follows: fetal skin, respiratory, urinary tract, gastrointestinal tract, and embryonic cells. This feature was first understood by Sancho and colleagues in 1993 when they obtained myoblasts from the amniotic fluid cells, and now is developed to a wide array of the cell types applied in tissue engineering [30].

Fig. 3.7 The schematic of different components of the womb and the placement of fetal stem cell sources, including placenta, amniotic fluid, and umbilical cord blood (Reprinted from [31], CC BY 4.0, modified from Gray's Anatomy)

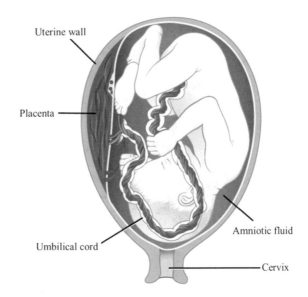

Uterine wall

Placenta

Amniotic fluid

Umbilical cord

Cervix

3.3.4 Adult Stem Cells

In each tissue, some stem cells named as adult stem cells play their role as a source of undifferentiated cells. By occurring any damage to a tissue, the cells support homeostasis and regeneration of the tissue. Adult stem cells are in the quiescence to preserve their stemness. Some sorts of adult stem cells are as follows: bone marrow, skin, muscle, and nervous system stem cells [16, 32]. Furthermore, mesenchymal stem cells obtained from the umbilical cords of newborn babies are also categorized as adult stem cells.

3.3.4.1 Bone Marrow Mesenchymal Stem Cells

Due to the role of mesenchymal stem cells in the natural process of bone formation, cell usage is greatly considered in tissue engineering since long time ago [2]. The cells inhabit in various locations such as blood, adipose tissue, skin, trabecular bone, and fetal blood, liver, and lung, as well as umbilical cord blood and placenta. Although many similar features are expected from mesenchymal stem cells, their potential in differentiation and gene expression is significantly different. Compared to embryonic stem cells and induced pluripotent stem cells, mesenchymal stem cells show some outstanding features as follows: (1) The highest osteogenic differentiation potential that maintains for a long time; (2) accessibility to autologous mesenchymal stem cells from a broad type of tissue; (3) feasible allogeneic type without immune rejection due to potential of cell to modifying immune responses [33]. As Fig. 3.8 shows, many kinds of cells can be differentiated from mesenchymal stem cells, including

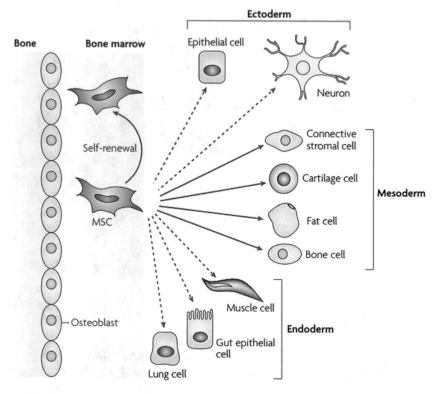

Fig. 3.8 The schematic of the ability of mesenchymal stem cells to proliferate and differentiate to various kinds of cells of different tissues (Reprinted from [35] by permission from Springer Nature)

osteoblasts, chondrocytes, myocytes, and adipocytes that attribute to different tissue consisted of the bone, cartilage, muscle, and fat, respectively. However, most of the mesenchymal stem cells are practically extracted from bone marrow and gingiva [34].

Bone marrow is the lake of various types of cells, named as stromal cells. Only lower than 0.01% of them is mesenchymal stem cells. However, the bone marrow-originated mesenchymal stem cells possess the highest multilineage potential compared to other kinds of mesenchymal stem cells, and it is considered greatly in tissue engineering application [33].

Weir and Xu [36] seeded human mesenchymal stem cells into a composite scaffold produced from calcium phosphate cement and chitosan. The scaffold can be used as a moderate load-bearing vehicle that carries human mesenchymal stem cells and enables their differentiation into the osteogenic lineage. In another study, calcium phosphate cement scaffolds in plain form or compositing with chitosan and vicryl fiber were fabricated. Both additives are degradable and can leave porosity in the composite structure. Human bone marrow mesenchymal stem cells were also added into the scaffolds after encapsulating in alginate hydrogel. The hydrogel preserved the cells from the harsh environment of the cement and improved osteogenesis ability of the composite [37]. Liu et al. [38] also, to survive mesenchymal stem cells during the cementation process, encapsulated the cells into chitosan/β-glycerophosphate.

The activity of two kinds of mesenchymal stem cells, one derived from human umbilical and the other from human bone marrow, was assessed by seeding on macroporous calcium phosphate cement and implanting the scaffolds on rat cranial defects. Comparing to the plain scaffold, bone formation and blood vessel density increased 57 and 15% for umbilical type, and 88 and 20% for bone marrow type, respectively [39]. In a study, besides bone marrow mesenchymal stem cells, platelet-rich plasma was also added to calcium phosphate cement scaffold, and the scaffolds were implanted in minipigs. In this scaffold, the ability of new bone production and vascularization was twice enhanced compared to naked scaffold [40].

3.3.4.2 Adipose Tissue-Derived Mesenchymal Stem Cells

Another kind of mesenchymal stem cell is the one derived from adipose tissue (as the constituents of the tissue are shown in Fig. 3.9). The cell kind is able to differentiate into bones and cartilage, and besides the bone marrow-originated kind is the most applicable mesenchymal stem cells in cell therapy [41]. The other features of adipose tissue-derived mesenchymal stem cells are their potential to differentiate into cells of mesodermal, ectodermal and, endodermal origin. Compared to bone marrow kind, these cells have the same marker on the surface, and similar proliferation and differentiation potential [42–44]. However, adipose-derived cells show some benefits related to bone marrow kind: (1) cause minimal discomfort for a patient during harvesting; (2) immune response can be much more modulated; and (3) the significantly greater potential of adipose tissue to produce mesenchymal stem cells [44, 45].

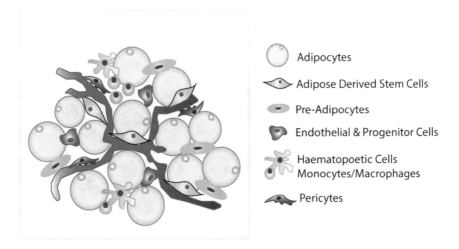

Fig. 3.9 The constituents of adipose tissue including mature adipocytes, preadipocytes, stem cells, endothelial cells, pericytes, macrophages, and circulating blood cells (Adapted from [46], CC BY)

3.3.4.3   Umbilical Cord Mesenchymal Stem Cells

Another source of mesenchymal stem cell is from the human umbilical cord. As shown in Fig. 3.10, the cord constructs an outer layer named amniotic epithelium, which covers a mucoid connective tissue called Wharton's Jelly. This tissue embeds three vessels, a vein and arteries. The overall length of the cord is about 60 cm with a diameter of 1.5 cm. Mesenchymal stem cell can be derived from each part of the cord [47, 48]. Among advantages of this cell kind, it can be mentioned to: (1) gene expression is in the same manner as embryonic stem cells and its proliferation is faster than bone marrow-derived one that makes it easy to obtain suitable number of mesenchymal stem cell; (2) the harvesting of the cells is noninvasive and ethical; (3) can be used in both autologous and allogeneic applications. Caution should be considered that the donor's health should be confirmed by genomic or chromosomal tests [49].

Using human umbilical cord mesenchymal stem cell is reported in many studies. Zhou et al. [50] by producing calcium phosphate cement-alginate microbead-vicryl fiber scaffold determined that an optimum cell seeding density is critical for suitable osteodifferentiation and mineralization. In another research reported that an increase in the volume fraction of poly(D, L-lactide-co-glycolide) fibers improved differentiation and mineralization of umbilical cord mesenchymal stem cell toward bone cells [51]. Chen et al. [52, 53] fabricated calcium phosphate cement-collagen scaffold via the gas foaming method and embedded alginate-fibrin microbeads containing the stem cells in it. The microbeads either protect the cells or release them during

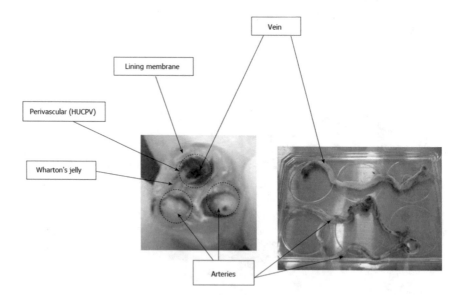

Fig. 3.10 The structure of umbilical cord (Reprinted from [49], CC BY-NC 4.0)

degradation. An increase in the collagen content also enhanced cellular activities. The same results was also obtained in another research [54].

3.3.4.4 Dental Tissues Derived Mesenchymal Stem Cells

Five kinds of mesenchymal stem cell originated from different parts of dental tissues, such as dental pulp, exfoliated deciduous teeth, periodontal ligament, apical papilla, and dental follicle (the parts are shown in Fig. 3.11). These cells can differentiate to either odontogenic cells or other cell lineages, but with lower potential compared to bone marrow originated ones [55].

Seeding dental follicle stem cells on tricalcium phosphate scaffold was evaluated. This study reported that tricalcium phosphate, even without the presence of an odontogenic inducer, can differentiate the cells to odontoblasts. In this system, the level of gene expression from differentiated odontoblasts was lower than the odontoblasts inhabited in pulp tissue [57]. In another research, the stem cells were seeded in a composite scaffold composed of calcium phosphate cement, chitosan, and metformin. The increase in the level of alkaline phosphatase and odontoblastic marker release besides mineral deposition indicated an enhancement of the odontogenic ability of the scaffold [58].

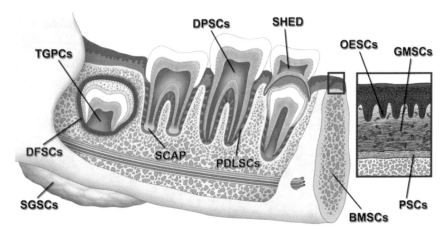

Fig. 3.11 The sources of mesenchymal stem cell originated from dental pulp (DPSCs), exfoliated deciduous teeth (SHED), periodontal ligament (PDLSCs), apical papilla (SCAP), and dental follicle (DFSCs) (Reprinted from [56] by permission from Elsevier)

3.3.4.5   Other Adult Stem Cells

Periosteum derived progenitor cells: A layer named periosteum (as shown in Fig. 3.12) covers the surface of the bone. The layer is constructed of two distinct layers, at which the external one is composed of elastic fibers and vessels and the inner layer named as cambium is the resource of periosteum-derived progenitor cells. In addition to osteogenesis applications, its ability for osteochondral treatment is also considered. Based on the reported researches, this cell type can also perform better than bone marrow-derived cells. Moreover, because of its high telomeres stability, the cell potential remains stable even after proliferation stages and its performance is similar to the young kind [59].

Synovium-derived mesenchymal stem cells: Another source of the mesenchymal stem cell is synovial membrane tissue that can be obtained from even a small amount harvested from the knee joint (Fig. 3.13). Besides the harvesting is done via arthroscopy, as minimally invasive surgery, the high productivity potential of the cell-embedded tissue enables full and fast healing process after surgery, as proven by in vivo investigations. The following lineages can be differentiated from synovial membrane tissue: osteocytes, adipocytes, and myocytes [60].

Muscle-derived stem cells: Skeletal muscle is another source for mesenchymal stem cell that can be obtained from only small biopsied part of the muscle. The cells show a high capacity to regenerate skeletal and cardiac muscle. The obtained cells also possess an acceptable potential to transduce by viral vectors and manipulate in the

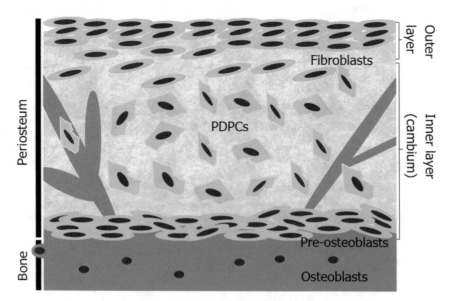

Fig. 3.12 Structure of the periosteum layer and its constitutions. PDPCs: periosteum-derived precursor cells (Reprinted from [59], CC BY-NC 4.0)

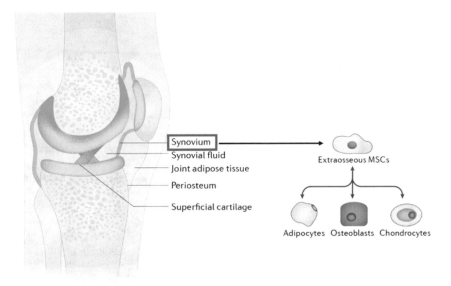

Fig. 3.13 The schematic of placement of synovial membrane tissue in the knee junction and the cells can be differentiated from the tissue (Adapted from [61] by permission from Springer Nature)

ex vivo situation. In genetically modified condition, the cells can differentiate into osteogenic and chondrogenic lineages [62].

3.3.5 Osteoblasts and Osteoclasts

3.3.5.1 Osteoblasts

One of the major types of bone cells that is responsible for bone matrix formation is osteoblast. It originates from mesenchymal osteoprogenitor cells derived from bone marrow and periosteum, and its secretory activity continues until apoptosis or changing phenotype to osteocytes and embeds in the bone matrix. The cell usually secretes collagen I, proteoglycans, glycoproteins, and ′γ-carboxylated proteins. Besides, it regulates osteoclast activities as a process to control the architecture of the bone. Irregularity in the activity of these two-type cells can lead to some diseases such as osteoporosis [63].

During osteoblast activities, bone-specific proteins and enzymes, such as alkaline phosphatase, osteocalcin, and collagen I, are released. After 3–6 weeks activities of the cells, they reach confluence, and lay a dense matrix with granular appearance containing calcium phosphate crystals. Unfortunately, the clinical use of osteoblasts faces with a drawback that is short lifespans of the cells [64].

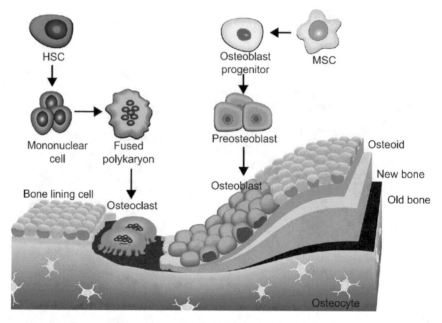

Fig. 3.14 The schematic of controlling bone homeostasis via activities of osteoclasts and osteoblasts (Reprinted from [66] by permission from Springer Nature)

3.3.5.2 Osteoclasts

Hematopoietic stem cell is the source of another kind of cell that is essential for bone homeostasis, named as osteoclast, the cell that resorbs bone (as schematically shown in Fig. 3.14). The stem cell creates mononuclear precursors of osteoclasts, i.e., preosteoclasts, and from the fusion of the obtained cells, multinucleated osteoclasts are formed. As mentioned, osteoblast can regulate osteoclast activities, but the reverse effect is not explicit [65].

In designing a scaffold, its degradation chemically and biologically should be considered to mimic natural bone remodeling and improve integration of the newly formed bone with the scaffold. In this manner, the scaffold should let osteogenic cell attachment and create a balance between the formation of new bone with osteoblasts and the resorption of the scaffold with osteoclasts. Studies show that although osteoclasts can resorb calcified matrixes such as calcium phosphate, tri-calcium phosphate, and calcite, the cells cannot erode poly(methyl methacrylate) [67].

3.4 Cell-Seeding Techniques

Cell seeding on a synthetic scaffold is a crucial procedure that determines the efficiency of the cell-to-cell contact and spreading cells in the scaffold. Due to the effect of these parameters on the success of the tissue engineering strategy, in this section, the method of cell seeding is discussed.

3.4.1 Static Cell Seeding Method

In the static cell seeding method, the cell suspension is simply seeded into a scaffold (as shown in Fig. 3.15). Although it is an easy method, it faces with great shortages including low cell density with an efficiency of 10–25% and insufficient cell distribution and penetration. Even its complete attachment cannot be assured [68].

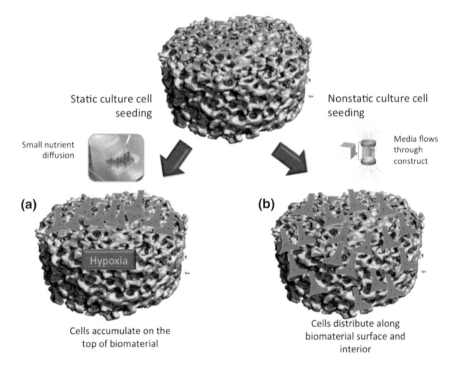

Fig. 3.15 The schematic illustration of **a** static cell seeding via direct pipetting onto a scaffold surface and **b** dynamic cell seeding that fluid and its biological constituents can be flow through a scaffold via applying rotation, vacuum, or magnetic forces (Reprinted from [69], CC BY 3.0)

3.4.2 Dynamic Cell Seeding

In the dynamic seeding method, a bioreactor is used in two manners, by applying rotation and benefits from hydrostatic forces or applying a pressure gradient. The efficiency of the methods is 38–90% and 60–90%, respectively. Some adverse effect was reported about the bad effect on the cell morphology during the application of the rotational method [68].

3.4.3 Magnetic Cell Seeding Method

In the magnetic cell seeding method, an external magnetic field improves the attachment of cells to a scaffold. The method has two modifications. In the first one, superparamagnetic monosized polymers, such as dynabeads, are initially embedded in a scaffold. Then by applying a magnetic field, the desired cells or proteins are attached to the scaffold. In the second method, cationic liposomes that contain superparamagnetic iron oxide particles first label cells and then by applying a magnetic force, the complex would be attached to the scaffold. The efficiency of the first and second methods is 99 and 90%, respectively. Although the methods are reproducible, their effects on cell and other tissues need more research [68].

3.5 Bioreactors

Although using a scaffold accompanied with seeded cells is essential to improve tissue regeneration in tissue engineering strategy, approaching to real condition needs more attempt. To this aim, some tissue-specific bioreactors, those involving biochemical and mechanical factors, should be applied to obtain cohesive, organized, and functional tissues on the scaffold. For example, as shown in Fig. 3.16, merely using dynamic bioreactors instead of static ones can improve the cell activity on a scaffold [70].

The expected features for a bioreactor are as follows: (1) enables to distribute cells uniformly on a scaffold; (2) involves essential condition for cell vitality including nutrients, gas atmosphere, and biological agents; (3) creates a dynamic flow for cell culture medium; (4) stimulates real physical conditions; and (5) facilitates system monitoring [71]. As shown in Fig. 3.17, bioreactors can be classified as spinner flask bioreactor, rotating wall bioreactor, hollow-fiber bioreactor, perfusion bioreactor, Compression bioreactor, and even their combinations.

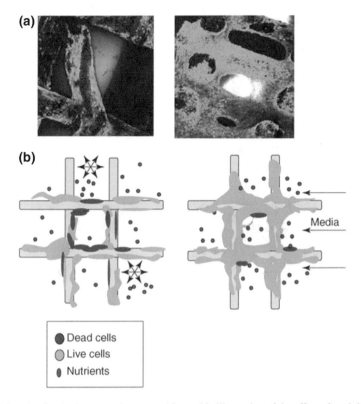

Fig. 3.16 **a** Confocal microscopy images and **b** graphic illustration of the effect of applying static (left) and dynamic (right) flow of culture medium on cell activity (Reprinted from [70] by permission from Elsevier)

3.5.1 Spinner Flask Bioreactor

A spinner flask is a bioreactor that with dynamic flow improves the proliferation and tissue regeneration of seeded cells on a scaffold. In this system (Fig. 3.17a), the fluid flows by the rotation of a magnetic stir bar and an impeller attached to the bar. The scaffold can be held stationary or be allowed to float freely. In the latter mode, lower shear is imposed on seeded cells. Dynamic flow improves the cell activity compared to static one, and controlling the imposed shear on cells can be applied via changing diameter, size, and rotation speed of the impeller. However, the shear stress should not be high that destroys cells [73].

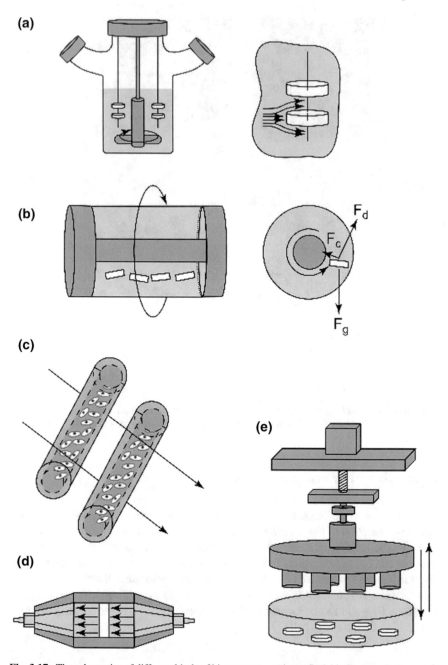

Fig. 3.17 The schematics of different kinds of bioreactors, **a** spinner flask bioreactor which creates flowing by a magnetic stir bar and attached impeller, **b** rotating wall bioreactor in which a scaffold can be maintained suspend via applying hydrodynamic drag force, **c** hollow fiber bioreactor which mimics vascular condition, **d** direct perfusion bioreactor that penetrating cell medium through scaffold by applying pump force, and **e** compression bioreactor that can stimulate mechanical loading conditions (Reprinted from [72] by permission from Elsevier)

3.5.2 Rotating Wall Bioreactor

Another dynamic bioreactor is the rotating wall type, developed by NASA to protect cell culture experiments during space missions. The reactor is constructed from a rotating wall that creates an upward hydrodynamic drag force to suspend a scaffold in the culture media (Fig. 3.17b). Like as spinner flask, cells distribute more homogeneously than in static reactors. The rotation speed of the wall is usually 15–30 rpm, and by growing tissue, it should be changed so that to maintain the suspension of the scaffold [74].

3.5.3 Hollow-Fiber Bioreactors

To mimic the in vivo structure of vessels, fibers are used as a reactor for seeded cells on a scaffold. In this system, fluid transfers across the fiber wall, and its rate does not directly affect the imposed shear stress on cells (Fig. 3.17c). Its high surface area to volume ratio can produce devices with less cell population variation and decrease the operating costs [75].

3.5.4 Perfusion Bioreactors

This bioreactor is composed of a pump, a culture media reservoir, a tubing circuit, and chambers to hold scaffolds. The pump perfuses the culture media through the scaffold, continuously or non- continuously (Fig. 3.17d). The scaffold should be highly porous, i.e., contains more than 70% porosity. Compared to the aforementioned bioreactors, perfusion bioreactor can create a more controlled and uniform flow that can be applied for large devices [76].

3.5.5 Compression Bioreactors

In a real situation, tissues are imposed with different kinds of mechanical forces. So, in some bioreactors, such as the compression bioreactor, applying compression forces via piston/compression systems or hydrodynamic compression are considered (Fig. 3.17e). Two critical issues should be considered, the one is an infection caused by the system that creates loads, and the other one is the ability of a scaffold to tolerate the applied loads [77].

3.5.6 Combined Bioreactors

To approach to a more mimicked bioreactor, some of the mentioned methods are combined, especially adding a perfusion loop is common, such as adding perfusion to the compression bioreactor. This system provides the condition of nutrient exchange [74].

References

1. O. Shinsuke, Y. Fumiko, C. Ung-il, Tissue engineering of bone and cartilage. IBMS Bonekey. **6**, 405–419 (2009)
2. A.R. Amini, C.T. Laurencin, S.P. Nukavarapu, Bone tissue engineering: recent advances and challenges. Crit. Rev. Biomed. Eng. **40**, 363–408 (2012). https://doi.org/10.1615/CritRevBiomedEng.v40.i5.10
3. I. Drosse, E. Volkmer, R. Capanna, P. De Biase, W. Mutschler, M. Schieker, Tissue engineering for bone defect healing: an update on a multi-component approach. Injury **39**, S9–S20 (2008). https://doi.org/10.1016/S0020-1383(08)70011-1
4. P.G. Robey, Cell sources for bone regeneration: the good, the bad, and the ugly (but promising). Tissue Eng. Part B Rev. **17**, 423–430 (2011). https://doi.org/10.1089/ten.teb.2011.0199
5. T. Yin, L. Li, The stem cell niches in bone. J. Clin. Invest. **116**, 1195–1201 (2006). https://doi.org/10.1172/JCI28568
6. D. Lopes, C. Martins-Cruz, M.B. Oliveira, J.F. Mano, Bone physiology as inspiration for tissue regenerative therapies. Biomaterials **185**, 240–275 (2018). https://doi.org/10.1016/j.biomaterials.2018.09.028
7. R. Nerem, A. Atala, R. Lanza, T. Mikos, Stem cells derived from amniotic fluid and placenta, in *Principles of Regenerative Medicine* (Academic Press, London, 2008), p. 1472
8. R. Florencio-Silva, G.R.D.S. Sasso, E. Sasso-Cerri, M.J. Simões, P.S. Cerri, Biology of bone tissue: structure, function, and factors that influence bone cells. Biomed. Res. Int. **2015**, 1–17 (2015). https://doi.org/10.1155/2015/421746
9. M. Capulli, R. Paone, N. Rucci, Osteoblast and osteocyte: games without frontiers. Arch. Biochem. Biophys. **561**, 3–12 (2014). https://doi.org/10.1016/j.abb.2014.05.003
10. K.A. Young, J.A. Wise, P. DeSaix, D.H. Kruse, B. Poe, E. Johnson, J.E. Johnson, O. Korol, J.G. Betts, M. Womble, *Anatomy and Physiology*, 1st edn. (Houston, Texas, USA, 2013)
11. B. Clarke, Normal bone anatomy and physiology. Clin. J. Am. Soc. Nephrol. **3**, S131–S139 (2008). https://doi.org/10.2215/CJN.04151206
12. K. Matsuo, Cross-talk among bone cells. Curr. Opin. Nephrol. Hypertens. **18**, 292–297 (2009). https://doi.org/10.1097/MNH.0b013e32832b75f1
13. F. Martini, J.L. Nath, E.F. Bartholomew, *Fundamentals of Anatomy & Physiology* (Essex, England, 2015)
14. M. Tang, W. Chen, M.D. Weir, W. Thein-Han, H.H.K. Xu, Human embryonic stem cell encapsulation in alginate microbeads in macroporous calcium phosphate cement for bone tissue engineering. Acta Biomater. **8**, 3436–3445 (2012). https://doi.org/10.1016/j.actbio.2012.05.016
15. M. Meregalli, A. Farini, Y. Torrente, Stem cell therapy for neuromuscular diseases, in *Stem Cells in Clinic and Research* (InTech, 2011). https://doi.org/10.5772/24013
16. S. Kargozar, M. Mozafari, S. Hamzehlou, P. Brouki Milan, H.-W. Kim, F. Baino, Bone tissue engineering using human cells: a comprehensive review on recent trends, current prospects, and recommendations. Appl. Sci. **9**, 174 (2019). https://doi.org/10.3390/app9010174

17. A.-M. Yousefi, P.F. James, R. Akbarzadeh, A. Subramanian, C. Flavin, H. Oudadesse, Prospect of stem cells in bone tissue engineering: a review. Stem Cells Int. **2016**, 1–13 (2016). https://doi.org/10.1155/2016/6180487

18. W. Chen, H. Zhou, M.D. Weir, M. Tang, C. Bao, H.H.K. Xu, Human embryonic stem cell-derived mesenchymal stem cell seeding on calcium phosphate cement-chitosan-RGD scaffold for bone repair. Tissue Eng. Part A **19**, 915–927 (2013). https://doi.org/10.1089/ten.tea.2012.0172

19. X. Liu, P. Wang, W. Chen, M.D. Weir, C. Bao, H.H.K. Xu, Human embryonic stem cells and macroporous calcium phosphate construct for bone regeneration in cranial defects in rats. Acta Biomater. **10**, 4484–4493 (2014). https://doi.org/10.1016/j.actbio.2014.06.027

20. L. Wang, P. Wang, M.D. Weir, M.A. Reynolds, L. Zhao, H.H.K. Xu, Hydrogel fibers encapsulating human stem cells in an injectable calcium phosphate scaffold for bone tissue engineering. Biomed. Mater. **11**, 065008 (2016). https://doi.org/10.1088/1748-6041/11/6/065008

21. Y. Lin, S. Huang, R. Zou, X. Gao, J. Ruan, M.D. Weir, M.A. Reynolds, W. Qin, X. Chang, H. Fu, H.H.K. Xu, Calcium phosphate cement scaffold with stem cell co-culture and prevascularization for dental and craniofacial bone tissue engineering. Dent. Mater. **35**, 1031–1041 (2019). https://doi.org/10.1016/j.dental.2019.04.009

22. S. Yamanaka, H.M. Blau, Nuclear reprogramming to a pluripotent state by three approaches. Nature **465**, 704–712 (2010). https://doi.org/10.1038/nature09229

23. J. Liu, W. Chen, Z. Zhao, H.H.K. Xu, Reprogramming of mesenchymal stem cells derived from iPSCs seeded on biofunctionalized calcium phosphate scaffold for bone engineering. Biomaterials **34**, 7862–7872 (2013). https://doi.org/10.1016/j.biomaterials.2013.07.029

24. M. Tang, W. Chen, J. Liu, M.D. Weir, L. Cheng, H.H.K. Xu, Human induced pluripotent stem cell-derived mesenchymal stem cell seeding on calcium phosphate scaffold for bone regeneration. Tissue Eng. Part A **20**, 1295–1305 (2014). https://doi.org/10.1089/ten.tea.2013.0211

25. W. TheinHan, J. Liu, M. Tang, W. Chen, L. Cheng, H.H.K. Xu, Induced pluripotent stem cell-derived mesenchymal stem cell seeding on biofunctionalized calcium phosphate cements. Bone Res. **1**, 371–384 (2013). https://doi.org/10.4248/BR201304008

26. P. Wang, X. Liu, L. Zhao, M.D. Weir, J. Sun, W. Chen, Y. Man, H.H.K. Xu, Bone tissue engineering via human induced pluripotent, umbilical cord and bone marrow mesenchymal stem cells in rat cranium. Acta Biomater. **18**, 236–248 (2015). https://doi.org/10.1016/j.actbio.2015.02.011

27. M. Sladkova, M. Palmer, C. Öhman, R.J. Alhaddad, A. Esmael, H. Engqvist, G.M. de Peppo, Fabrication of macroporous cement scaffolds using PEG particles: in vitro evaluation with induced pluripotent stem cell-derived mesenchymal progenitors. Mater. Sci. Eng., C **69**, 640–652 (2016). https://doi.org/10.1016/j.msec.2016.06.075

28. M. Sladkova, M. Palmer, C. Öhman, J. Cheng, S. Al-Ansari, M. Saad, H. Engqvist, G.M. de Peppo, Engineering human bone grafts with new macroporous calcium phosphate cement scaffolds. J. Tissue Eng. Regen. Med. **12**, 715–726 (2018). https://doi.org/10.1002/term.2491

29. X. Liu, W. Chen, C. Zhang, W. Thein-Han, K. Hu, M.A. Reynolds, C. Bao, P. Wang, L. Zhao, H.H.K. Xu, Co-seeding human endothelial cells with human-induced pluripotent stem cell-derived mesenchymal stem cells on calcium phosphate scaffold enhances osteogenesis and vascularization in rats. Tissue Eng. Part A **23**, 546–555 (2017). https://doi.org/10.1089/ten.tea.2016.0485

30. R.P. Dorin, C.J. Koh, Fetal tissues, in *Principles of Regenerative Medicine* (Elsevier, 2011), pp. 819–832. https://doi.org/10.1016/b978-0-12-381422-7.10045-8

31. C. Molnar, J. Gair, Human pregnancy and birth, in *Concepts of Biology*, 1st edn. (Houston, Texas, USA, 2013)

32. A.J. Wagers, I.L. Weissman, Plasticity of adult stem cells. Cell **116**, 639–648 (2004). https://doi.org/10.1016/S0092-8674(04)00208-9

33. A.H. Undale, J.J. Westendorf, M.J. Yaszemski, S. Khosla, Mesenchymal stem cells for bone repair and metabolic bone diseases. Mayo Clin. Proc. **84**, 893–902 (2009). https://doi.org/10.4065/84.10.893

34. H. Xia, X. Li, W. Gao, X. Fu, R.H. Fang, L. Zhang, K. Zhang, Tissue repair and regeneration with endogenous stem cells. Nat. Rev. Mater. **3**, 174–193 (2018). https://doi.org/10.1038/s41578-018-0027-6
35. A. Uccelli, L. Moretta, V. Pistoia, Mesenchymal stem cells in health and disease. Nat. Rev. Immunol. **8**, 726–736 (2008). https://doi.org/10.1038/nri2395
36. M.D. Weir, H.H.K. Xu, Culture human mesenchymal stem cells with calcium phosphate cement scaffolds for bone repair, J. Biomed. Mater. Res. Part B Appl. Biomater. **93B**, 93–105 (2010). https://doi.org/10.1002/jbm.b.31563
37. M.D. Weir, H.H.K. Xu, Human bone marrow stem cell-encapsulating calcium phosphate scaffolds for bone repair. Acta Biomater. **6**, 4118–4126 (2010). https://doi.org/10.1016/j.actbio.2010.04.029
38. T. Liu, J. Li, Z. Shao, K. Ma, Z. Zhang, B. Wang, Y. Zhang, Encapsulation of mesenchymal stem cells in chitosan/β-glycerophosphate hydrogel for seeding on a novel calcium phosphate cement scaffold. Med. Eng. Phys. **56**, 9–15 (2018). https://doi.org/10.1016/j.medengphy.2018.03.003
39. W. Chen, J. Liu, N. Manuchehrabadi, M.D. Weir, Z. Zhu, H.H.K. Xu, Umbilical cord and bone marrow mesenchymal stem cell seeding on macroporous calcium phosphate for bone regeneration in rat cranial defects. Biomaterials **34**, 9917–9925 (2013). https://doi.org/10.1016/j.biomaterials.2013.09.002
40. G. Qiu, Z. Shi, H.H.K. Xu, B. Yang, M.D. Weir, G. Li, Y. Song, J. Wang, K. Hu, P. Wang, L. Zhao, Bone regeneration in minipigs via calcium phosphate cement scaffold delivering autologous bone marrow mesenchymal stem cells and platelet-rich plasma. J. Tissue Eng. Regen. Med. **12**, e937–e948 (2018). https://doi.org/10.1002/term.2416
41. J. Pak, J.H. Lee, K.S. Park, M. Park, L.-W. Kang, S.H. Lee, Current use of autologous adipose tissue-derived stromal vascular fraction cells for orthopedic applications. J. Biomed. Sci. **24**, 9 (2017). https://doi.org/10.1186/s12929-017-0318-z
42. S. Kern, H. Eichler, J. Stoeve, H. Klüter, K. Bieback, Comparative analysis of mesenchymal stem cells from bone marrow, umbilical cord blood, or adipose tissue. Stem Cells **24**, 1294–1301 (2006). https://doi.org/10.1634/stemcells.2005-0342
43. Y. Zhu, T. Liu, K. Song, X. Fan, X. Ma, Z. Cui, Adipose-derived stem cell: a better stem cell than BMSC. Cell Biochem. Funct. **26**, 664–675 (2008). https://doi.org/10.1002/cbf.1488
44. Y. Alabdulkarim, B. Ghalimah, M. Al-Otaibi, H. Al-Jallad, M. Mekhael, B. Willie, R. Hamdy, Recent advances in bone regeneration: the role of adipose tissue-derived stromal vascular fraction and mesenchymal stem cells. J. Limb Lengthening Reconstr. **3**, 4 (2017). https://doi.org/10.4103/jllr.jllr_1_17
45. R.T. Qomi, M. Sheykhhasan, Adipose-derived stromal cell in regenerative medicine: a review. World J. Stem Cells **9**, 107 (2017). https://doi.org/10.4252/wjsc.v9.i8.107
46. L. Shukla, W.A. Morrison, R. Shayan, Adipose-derived stem cells in radiotherapy injury: a new frontier. Front. Surg. **2** (2015). https://doi.org/10.3389/fsurg.2015.00001
47. J.T. Walker, A. Keating, J.E. Davies, Stem cells: umbilical cord/Wharton's jelly derived, in *Cell Engineering and Regeneration* (Springer International Publishing, Cham, 2019), pp. 1–28. https://doi.org/10.1007/978-3-319-37076-7_10-1
48. D.-C. Ding, Y.-H. Chang, W.-C. Shyu, S.-Z. Lin, Human umbilical cord mesenchymal stem cells: a new era for stem cell therapy. Cell Transplant. **24**, 339–347 (2015). https://doi.org/10.3727/096368915X686841
49. T. Nagamura-Inoue, Umbilical cord-derived mesenchymal stem cells: their advantages and potential clinical utility. World J. Stem Cells. **6**, 195 (2014). https://doi.org/10.4252/wjsc.v6.i2.195
50. H. Zhou, M.D. Weir, H.H.K. Xu, Effect of cell seeding density on proliferation and osteodifferentiation of umbilical cord stem cells on calcium phosphate cement-fiber scaffold. Tissue Eng. Part A **17**, 2603–2613 (2011). https://doi.org/10.1089/ten.tea.2011.0048
51. C. Bao, W. Chen, M.D. Weir, W. Thein-Han, H.H.K. Xu, Effects of electrospun submicron fibers in calcium phosphate cement scaffold on mechanical properties and osteogenic differentiation of umbilical cord stem cells. Acta Biomater. **7**, 4037–4044 (2011). https://doi.org/10.1016/j.actbio.2011.06.046

52. W. Chen, H. Zhou, M. Tang, M.D. Weir, C. Bao, H.H.K. Xu, Gas-foaming calcium phosphate cement scaffold encapsulating human umbilical cord stem cells. Tissue Eng. Part A **18**, 816–827 (2012). https://doi.org/10.1089/ten.tea.2011.0267

53. W. Chen, H. Zhou, M.D. Weir, C. Bao, H.H.K. Xu, Umbilical cord stem cells released from alginate–fibrin microbeads inside macroporous and biofunctionalized calcium phosphate cement for bone regeneration. Acta Biomater. **8**, 2297–2306 (2012). https://doi.org/10.1016/j.actbio.2012.02.021

54. W. Thein-Han, H.H.K. Xu, Collagen-calcium phosphate cement scaffolds seeded with umbilical cord stem cells for bone tissue engineering. Tissue Eng. Part A **17**, 2943–2954 (2011). https://doi.org/10.1089/ten.tea.2010.0674

55. G.T.-J. Huang, S. Gronthos, S. Shi, Mesenchymal stem cells derived from dental tissues vs. those from other sources: their biology and role in regenerative medicine. J. Dent. Res. **88**, 792–806 (2009). https://doi.org/10.1177/0022034509340867

56. H. Egusa, W. Sonoyama, M. Nishimura, I. Atsuta, K. Akiyama, Stem cells in dentistry—Part I: stem cell sources. J. Prosthodont. Res. **56**, 151–165 (2012). https://doi.org/10.1016/j.jpor.2012.06.001

57. M.B. Eslaminejad, S. Bordbar, H. Nazarian, Odontogenic differentiation of dental pulp-derived stem cells on tricalcium phosphate scaffolds. J. Dent. Sci. **8**, 306–313 (2013). https://doi.org/10.1016/j.jds.2013.03.005

58. W. Qin, J.-Y. Chen, J. Guo, T. Ma, M.D. Weir, D. Guo, Y. Shu, Z.-M. Lin, A. Schneider, H.H.K. Xu, Novel calcium phosphate cement with metformin-loaded chitosan for odontogenic differentiation of human dental pulp cells. Stem Cells Int. **2018**, 1–10 (2018). https://doi.org/10.1155/2018/7173481

59. C. Ferretti, Periosteum derived stem cells for regenerative medicine proposals: boosting current knowledge. World J. Stem Cells. **6**, 266 (2014). https://doi.org/10.4252/wjsc.v6.i3.266

60. J. Fan, R.R. Varshney, L. Ren, D. Cai, D.-A. Wang, Synovium-derived mesenchymal stem cells: a new cell source for musculoskeletal regeneration. Tissue Eng. Part B Rev. **15**, 75–86 (2009). https://doi.org/10.1089/ten.teb.2008.0586

61. D. McGonagle, T.G. Baboolal, E. Jones, Native joint-resident mesenchymal stem cells for cartilage repair in osteoarthritis. Nat. Rev. Rheumatol. **13**, 719–730 (2017). https://doi.org/10.1038/nrrheum.2017.182

62. A. Usas, J. Huard, Muscle-derived stem cells for tissue engineering and regenerative therapy **28**, 5401–5406 (2007)

63. E.J. Mackie, Osteoblasts: novel roles in orchestration of skeletal architecture. Int. J. Biochem. Cell Biol. **35**, 1301–1305 (2003). https://doi.org/10.1016/S1357-2725(03)00107-9

64. P. Jayakumar, L. Di Silvio, Osteoblasts in bone tissue engineering. Proc. Inst. Mech. Eng. Part H J. Eng. Med. **224**, 1415–1440 (2010). https://doi.org/10.1243/09544119jeim821

65. D. Han, Q. Zhang, An essential requirement for osteoclasts in refined bone-like tissue reconstruction in vitro. Med. Hypotheses **67**, 75–78 (2006). https://doi.org/10.1016/j.mehy.2006.01.014

66. M.S. Rahman, N. Akhtar, H.M. Jamil, R.S. Banik, S.M. Asaduzzaman, TGF-β/BMP signaling and other molecular events: regulation of osteoblastogenesis and bone formation. Bone Res. **3**, 15005 (2015). https://doi.org/10.1038/boneres.2015.5

67. S. Midha, W. van den Bergh, T.B. Kim, P.D. Lee, J.R. Jones, C.A. Mitchell, Bioactive glass foam scaffolds are remodelled by osteoclasts and support the formation of mineralized matrix and vascular networks in vitro. Adv. Healthc. Mater. **2**, 490–499 (2013). https://doi.org/10.1002/adhm.201200140

68. M. Baghaban, F. Faghihi, Mesenchymal stem cell-based bone engineering for bone regeneration, in *Regenerative Medicine and Tissue Engineering—Cells and Biomaterials* (InTech, 2011). https://doi.org/10.5772/21017

69. J. Lott, P.H. de Carvalho, D. Assis, A.M. de Goes, Innovative strategies for tissue engineering, in *Advances in Biomaterials Science and Biomedical Applications* (InTech, 2013). https://doi.org/10.5772/53337

70. D.W. Hutmacher, H. Singh, Computational fluid dynamics for improved bioreactor design and 3D culture. Trends Biotechnol. **26**, 166–172 (2008). https://doi.org/10.1016/j.tibtech.2007.11.012
71. M. Sladkova, G. de Peppo, Bioreactor systems for human bone tissue engineering. Processes **2**, 494–525 (2014). https://doi.org/10.3390/pr2020494
72. I. Martin, D. Wendt, M. Heberer, The role of bioreactors in tissue engineering. Trends Biotechnol. **22**, 80–86 (2004). https://doi.org/10.1016/j.tibtech.2003.12.001
73. K.J. Blose, J.T. Krawiec, J.S. Weinbaum, D.A. Vorp, Bioreactors for tissue engineering purposes, in *Regenerative Medicine Applications in Organ Transplantation* (Elsevier, 2014), pp. 177–185. https://doi.org/10.1016/b978-0-12-398523-1.00013-6
74. N. Plunkett, F.J. O'Brien, IV.3. Bioreactors in tissue engineering. Stud. Health Technol. Inform. **152**, 214–230 (2010)
75. N. Wung, S.M. Acott, D. Tosh, M.J. Ellis, Hollow fibre membrane bioreactors for tissue engineering applications. Biotechnol. Lett. **36**, 2357–2366 (2014). https://doi.org/10.1007/s10529-014-1619-x
76. D.A. Gaspar, V. Gomide, F.J. Monteiro, The role of perfusion bioreactors in bone tissue engineering. Biomatter **2**, 167–175 (2012). https://doi.org/10.4161/biom.22170
77. A.J. El Haj, S.H. Cartmell, Bioreactors for bone tissue engineering. Proc. Inst. Mech. Eng. Part H J. Eng. Med. **224**, 1523–1532 (2010). https://doi.org/10.1243/09544119jeim802

Chapter 4
Inductivity: Bioactive Agents

4.1 Introduction

Besides the conductivity and productivity, inductivity is also crucial in tissue engineering. In bone tissue engineering, the process, called the osteoinductivity, is usually mimicked from natural bone healing and remodeling. In general, during osteoinductivity, immature cells are induced to differentiate into preosteoblasts to develop the osteogenesis process [1]. The induction is controlled by factors and proteins that reside in the extracellular matrix of bone. By binding the factors to cell surface receptors, the configuration of the receptors changes, and some signals are transferred into the intracellular environment. The signal activates a protein kinase that transcripts messenger RNA into an intra- or extracellular activated protein. The cumulation and activity of the proteins regulate the bone healing and remodeling process [2].

The factors are generally categorized into the following classes: transforming growth factor (TGF-β), insulin-like growth factor (IGF, I and II), platelet-derived growth factor (PDGF), fibroblast growth factor (FGF), and the bone morphogenetic proteins (BMPs). The researches, performed in recent decades, are focused on isolating and synthesis the factors, and in consequence, studying their activities alone and in combination to understand the role of each factor in the osteogenesis process. The works have resulted in improving the osteogenesis potential of bone grafts and some of them are even introduced in the market [2].

In bone tissue engineering, a scaffold is osteoinductive that can (1) carry or activate mesenchymal stem cells, (2) mature the cells into osteoblasts, and (3) improve new bone formation [3]. Because many studies have focused on the combination of scaffolds and bioactive agents [4], in this chapter, the agents are introduced, and their applications in bone tissue engineering of the bone cement scaffold are reviewed.

© The Author(s), under exclusive license to Springer Nature Switzerland AG 2020
H. Reza Rezaie et al., *Bone Cement*, SpringerBriefs in Applied
Sciences and Technology, https://doi.org/10.1007/978-3-030-39716-6_4

4.2 Growth Factors Involved in Bone Regeneration

After bone breaking and during bone remodeling, which regularly takes place, a complex process of activation of various cells and bioactive agents happens to activate, proliferate, differentiate, and migrate the osteogenesis cells. The ability to understand and control the process is the golden key to determine and define a comprehensive strategy in the bone tissue engineering [5]. As shown in Fig. 4.1, the process is divided into three stages, including inflammation and vascularization, soft callus formation, and hard callus formation, and continued with a complementary stage called the remodeling. In the first stage, instantly after occurring injury, an obstacle is created by blood clotting to prevent contacting the inner part of the body with the exterior environment. In this stage, many inflammatory cells, fibroblasts, and mesenchymal stem cells become activated alongside the release of various growth factors and cytokines to appear non-infectious inflammatory response and vascularization. In the second stage, the formation of osteoblasts and chondrocytes from progenitor cells happens and the granulation tissue, formed in the previous stage, is gradually substituted by fibro-cartilage tissue. The last stage is completed by mineralization of soft callus and the formation of hard callus from the woven bone. Osteoclasts and osteoblasts activities continue the healing process by replacing the hard callus with secondary lamellar containing a normal vascular supply [4].

Besides the mentioned cells, the molecules that participate in signaling during the bone healing process can be listed as follows: inflammatory growth factors, angiogenic growth factors, osteogenic growth factors, and systemic factors.

Inflammatory growth factors: Some proinflammatory cytokines and growth factors, that in inflammation stage, play their roles to stimulate the migration and differentiation of osteoblasts and osteoclasts, and also activate the secondary signal cascade for the next stage, are listed as follows: tumor necrosis factor-α (TNF-α), interleukins (ILs), interferon-γ (IFN-γ) and prostaglandins (PGs) [5].

Angiogenic growth factors: To supply required nutrients for cells that are active in the bone healing process, a vast network of vessels is crucial. The vascularization that lays through different stages of the bone healing process controls with the following factors: vascular endothelial growth factor (VEGF), platelet-derived growth factor (PDGFs), angiopoietin (Angs), FGFs, and IGFs [4].

Osteogenic growth factors: A superfamily called TGF-β with 60 members, including BMPs, activins, TGF-βs, and growth differentiation factor (GDF), is responsible for osteogenesis. The factors differentiate aggregated mesenchymal stem cells into chondrocytes or osteoblasts and continue the bone healing process through endochondral ossification and intramembranous ossification, respectively [4].

Systemic factors: During bone healing, some factors that usually control homeostasis of bone are also involved, such as parathyroid hormone (PTH), growth hormone, steroids, calcitonin, and vitamin D. Among them specially PTH and calcitonin are known due to their effect on regulating the calcium balance [4].

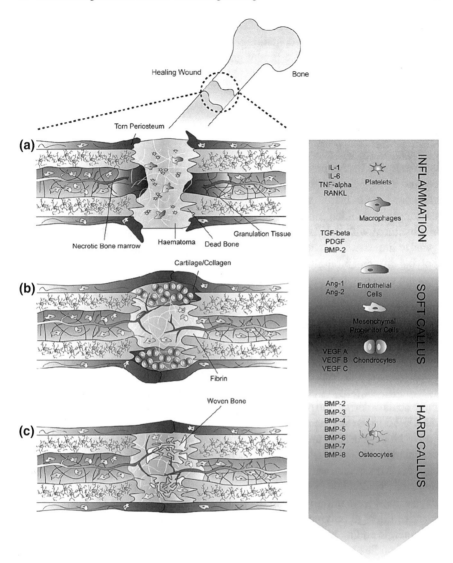

Fig. 4.1 The schematic of the bone healing process in three main stages of **a** inflammatory, **b** soft callus, and **c** hard callus phases and accompanying of various cells and bioactive agents (Reprinted from [6] by permission from Elsevier)

4.3 Bioactive Agent Used in Bone Tissue Engineering

In general, bone tissue is poorly vascularized. Hence, treating bone diseases, such as osteoporosis and osteonecrosis, with general drug administration, is a challenge. The issue becomes more severe when this vascularization is disrupted by trauma or during a surgical process. The mentioned problems are the main reason to encourage scientist to utilize implants containing biological agents and compensate for the shortage [7]. In this manner, different growth factors, bioactive biomolecules, and drugs have been studied and are discussed in the following sections.

4.3.1 Growth Factors

As mentioned in the previous chapter, in bone tissue engineering, a scaffold is required to support growing new cells and tissue. The scaffold should have appropriate porosity to facilitate vascularization and tissue growth through its structure, and supply demanded nutrients. In addition, the scaffold should be degradable to be substituted with new bone tissue and also enable the release of some biological agents to improve osteogenesis or angiogenesis [8]. Hence in the following sections, applying different biological agents in bone tissue engineering is discussed, and some common uses of them are listed in Table 4.1.

4.3.1.1 Bone Morphogenetic Proteins (BMPs)

Seven of 20 identified members of the BMP family are confirmed that they have osteoinductive potential. BMPs are composed of amino acid sequences and two disulfide-linked polypeptide subunits that weight 30–38 kDa, in general. Based on amino acids ordering, the proteins are divided into two main groups, BMP-2/-4 group and BMP-5 to -8 (also named as osteogenic protein-1 or OP-1 group). Although the diverse kinds of BMPs are present, only commercially available ones are focused, particularly BMP-2 and BMP-7 that have been proved clinically [9]. The heterodimer usage of BMPs is also evaluated and shows improvement in their application potential compared to individual usage [10]. For example, as shown in Fig. 4.2, hydroxyapatite nanoparticles were initially decorated with polydopamine and BMP-2 to be used as bioactive fillers in polymethyl methacrylate cement. The in vitro results confirmed that synergic adding bioactive agents enhance the bone regeneration [11].

Many studies have been conducted on the application of BMPs in bone cement. Zeng et al. [13] applied allogeneic bone powder and β-tricalcium phosphate incorporated with BMP to improve biocompatibility and osteogenic induction ability of calcium phosphate cement. Besides this kind of cement, brushite [14], Calcium sulfate-HA [15], calcium phosphate cement [16], magnesium-doped calcium phosphate cement [17], strontium-doped calcium phosphate cement [18], and chondroitin sulfate-functionalized calcium phosphate cement [19] were also evaluated. Calcium phosphate cement in plain mode and loaded with BMP-2 were compared and confirmed the high osteogenesis potential of the latter mode [20]. Baek et al. [21] showed (Fig. 4.3) that by adding BMP into hydroxyapatite microspheres, more new bone tissue grew in the implanted site.

Table 4.1 Growth factors practiced commonly in bone tissue engineering (Reprinted from [12] by permission from Springer Nature)

Growth factor	Cell source	Biologic effect	Action on bone
BMP	Osteoprogenitor cell, osteoblast, chondrocyte, endothelial cell (BMP-2)	Chondro-osteogenesis, osteoinduction (BMP-2)	Migration of osteoprogenitors, induction of proliferation, differentiation and matrix synthesis
FGF	Macrophage, monocyte, BMSC, chondrocyte, osteoblast, endothelial cell	Angiogenesis, proliferation of fibroblast and smooth muscle cells of vessels	Chondrocyte maturation (FGF-1). Osteoblast proliferation and differentiation, inhabitation of apoptosis of immature osteoclasts, induction of apoptosis of mature osteocytes, bone resorption (FGF-2)
IGF	Osteoblast, chondrocyte, hepatocyte, endothelial cell	Regulation of growth hormone effects	Osteoblast proliferation and bone matrix synthesis, bone resorption
PDGF	Platelet, osteoblast, endothelial cell, monocyte, macrophage	Proliferation of connective tissue cells, monocyte/macrophage and smooth muscle cell chemotaxis, angiogenesis	Osteoprogenitor migration, proliferation and differentiation
TGF-β	Platelet, osteoblast, BMSC, chondrocyte, endothelial cell, fibroblast, macrophage	Immunosuppression, angiogenesis, stimulation of cell growth, differentiation and ECM synthesis	Undifferentiated mesenchymal cell proliferation, osteoblast precursor recruiting; osteoblast and chondrocyte differentiation (but inhibition of terminal differentiation), bone matrix production, recruitment of osteoclast precursors
VEGF	Osteoblast, platelet	Angiogenesis	Conversion of cartilage into bone, osteoblast proliferation and differentiation

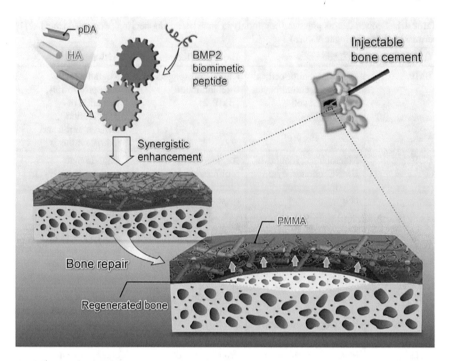

Fig. 4.2 Improving bioactivity of polymethyl methacrylate by adding polydopamine and BMP-2-coated hydroxyapatite particles (Reprinted from [11] by permission from Elsevier)

In a comparative study, some commercial biomaterials, such as β-tricalcium phosphate, calcium phosphate cement, and polylactic/polyglycolic acid, incorporated with BMPs, were assessed. The highest bone regeneration was observed in β-tricalcium phosphate kind. Most of calcium phosphate cement remained because of its low degradability. In the polylactic/polyglycolic acid, although the materials wholly degraded, the remained polymer fragments created inflammatory response besides bone regeneration [22].

4.3.1.2   Transforming Growth Factor-Beta (TGF-β)

One of the factor groups that appears in various cell and tissue activities is TGF-βs, with three common isoforms in the human body including TGF-β1 through -β3. In general, the factors are sourced from the extracellular bone matrix in the form of an inactive complex with latency-associated peptide, and also from platelets in the blood clot. The roles of TGF-βs in osteogenesis are stimulation of osteoprogenitor cell migration and regulation of proliferation, differentiation, and extracellular matrix synthesis. However, in different research performed on the factors, both the stimulatory and inhibitory effects on bone formation have been observed [10].

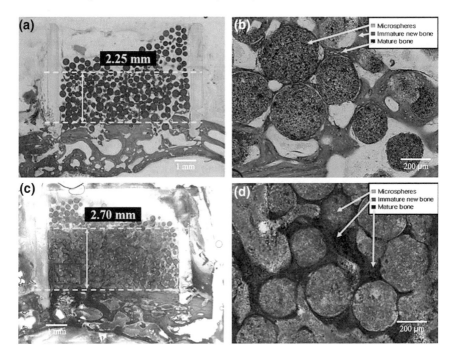

Fig. 4.3 Histological evaluation of two implanted hydroxyapatite microspheres containing **a** and **b** no biological agents, and **c** and **d** BMP, after 4 weeks. The images show that newly grown bone appears more in the c and d parts (Reprinted from [21] by permission from Elsevier)

Many studies have evaluated adding TGF-β to cement. Loading 0.75 mg TGF-β1 in calcium phosphate cement was not effective on osteogenesis of the cement [23]. However, adding poly(lactic-co-glycolic acid) containing 200 ng TGF-β1 to the cement improved the growth of new bone tissue significantly after 8-week implanting in a rat model [24]. Furthermore, Link et al. [25] loaded gelatin microparticles with TGF-β1 and added them into calcium phosphate cement. The presence of the factor enhanced degradation of the composite after 12-week implantation in the femoral condyles of rabbits.

4.3.1.3 Fibroblast Growth Factor (FGF)

Although FGF-2 stimulates the bone healing process, its performance depends on the practicing dose and exposure time. Because of this dependence in some studies, osteogenesis inhabitation was reported, and in the others sustains proliferation and osteogenesis. The occurrence originates from the ability of the FGF family in stimulating cell proliferation that can increase the osteoclast activity and create an imbalance in the presence of BMP antagonist noggin and BMP [26].

The presence of FGF-2 in calcium phosphate cement compared to the plain one indicated a 90% increase in osteogenesis [27]. In another research that studied osteogenesis of (poly(2-hydroxyethyl methacrylate/trimethylolpropane trimethacrylate)), as a dental resin cement, reported that adding polyHEMA/TMPT particles containing FGF-2 improved bone regeneration compared to the plain cement [28].

4.3.1.4   Insulin-Like Growth Factor (IGF)

IGF mediates the function of hormone, growth factors, cytokines, and morphogens that are active in the bone healing process in different inductive molecular levels, containing endocrine, paracrine, and autocrine levels. The most crucial roles of IGF are stimulating the proliferation and chemotactic migration of various cells and to affect the bone metabolism. Because of it, during bone formation and mineralization, many studies reported the release of IGF and its effect on osteogenesis [26]. For example, Vahabzadeh et al. [29] reported enhancement of bone regeneration when compared silicon/zinc-doped brushite cement containing IGF with a plain one.

4.3.1.5   Vascular Growth Factor (VEGF)

Vascularization cannot be abandoned during osteogenesis because of its vital effects on supplying nutrients and paracrine factors. VEGF is a factor that accomplishes vascularization via inducing proliferation and migration of endothelial cells. In addition to the mentioned task, the factor is effective on recruitment, survival, and activity of bone forming cells by two-way mediation of TGF-β1, IGF, and FGF-2. The performance of VEGF has encouraged scientists to accompany the factor with osteoinductive factors to achieve both osteogenesis and angiogenesis [26]. For instance, Lode et al. [30, 31] improved osteogenesis feature of calcium phosphate cement via adding VEGF. In another study, calcium phosphate cement was loaded with chitosan/dextran sulfate microparticles embedded VEGF. The encapsulation strategy was performed to preserve the osteogenesis and angiogenesis properties of the scaffold [32]. Other investigations also considered vascularization feature of scaffold composed from calcium phosphate cement containing VEGF [33, 34].

4.3.1.6   Platelet-Derived Growth Factor (PDGF)

PDGF indirectly mediates the fibroblast and osteoblasts activity via inducing bone, cartilage, and vascular mesenchymal stem cells. The factor is a glycolytic protein with the two disulfide-linked peptide chains (A and B) that create various kinds of PDGF, such as PDGF-AA, PDGF-BB, and PDGF-AB. The factors usually release from platelets [26].

Table 4.2 Drawbacks of applying single factor in tissue engineering (Reprinted from [35] by permission from Elsevier)

	Drawbacks	Implications
Single growth factor	• Control of cellular behavior	• Inadequate control of: migration, differentiation, morphogenesis, apoptosis
		• Inability to mimic dynamics of physiological cell signal controls
	• Co-activation/priming	• Reduced bioactivity of cells due to insufficient activation and crosstalk
		• Receptor dimerization

4.3.1.7 Delivery of Multi Growth Factors

Due to the limitation of delivery of single growth factor (as mentioned in Table 4.2), co-delivery of factors gets much attention. Many types of research attempt to use a combinational and sequential release of factors, but some challenges are faced with the improvement. The challenges are as follows: the selection of growth factors, the interaction of them, and the effect of dose, gradient, and release timing. For example, in Fig. 4.4, the schematic illustrates the real bone healing process, and step by step shows how a combination strategy should be considered to mimic the real condition. As shown, the considered conditions are variation in each growth factor concentration with time, the distance between each release peak, and the overlap area of two factors [35]. Table 4.3 also summarizes the studies benefit from using multi growth factors.

4.3.2 Drugs

The goals encouraged scientists to use active molecules include enhancing the potential of bone regeneration and targeting specific diseases [43]. Types of cement, for example, calcium phosphate cement, inherently can be applied as drug carriers. This potential originates from their setting at room temperature and their ability to be injected. Various drugs, such as antibiotics, anti-osteoporotic, anti-inflammatory, and anticancer drugs, are incorporated in bone cement [7], and some of them are summarized in Table 4.4.

4.3.2.1 Antibiotics

Wide usage of antibiotics originates from their two critical applications including preventing from and treating infections. The main challenge after surgical implantation

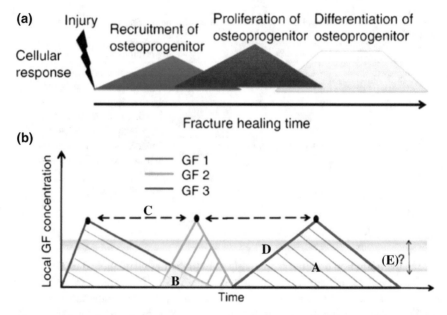

Fig. 4.4 a The schematic of creating an injury in bone and three different phases happen during the bone healing process. **b** Releasing multiple growth factors with considering their concentration **A**, the peak of release, and timing. Besides, the release of multi factors can happen with an overlap **B** or in consequence, with time distance **C** based on interaction among growth factors. The rate of release **D** can be determined concerning in-vitro evaluations. The effective dosage **E** for factor release that outrange release can create an inappropriate effect (Adapted from [35] by permission from Elsevier)

Table 4.3 Various studies on using multi growth factors

Growth factors	Carrier	Cement	Ref.
TGF-β1 and BMP-2	Ti-fiber	CPC	[36]
PDGF to BMP-2	PLGA microsphere and Alginate hydrogel	CPC	[37]
PDGF, VEGF and TGF-β1	PLGA microsphere	brushite	[38]
PDGF and IGF	PLGA microsphere	CPC	[39, 40]
TGF-β1 and VEGF		HA and calcium sulfate hemihydrate	[41]
TGF-β1 and BMP-2	Alginate hydrogel	CPC	[42]

is to prevent infection. Infection can fail the performed surgery and create many problems. Usually, antibiotic treatment performs either orally or intravenously. However, the inability to easily access to the demanded site may lengthen the treatment period. Hence, direct usage of antibiotic in cement can considerably solve the issue [44].

Table 4.4 Some studied drugs in musculoskeletal disorders (Reprinted from [8] by permission from Elsevier)

Name	Type
Gentamicin	Antibiotic
Vancomycin	
Tetracycline	
Alendronate	A member of bisphosphonate drug, used to treat bone diseases like osteoporosis
Zoledronate	
Human lactoferrin (hLFI-II)	Antimicrobial peptide (AMP)
Cephalexin	Semisynthetic cephalosporin antibacterial agent
Methotrexate	Anticancer agent
Cis-Platin	
Ceramide	
Doxycycline hyclate (DOXY-h)	Antibiotic, commonly used in dentistry to defeat periodontal pathogens
Ibuprofen	A non-steroidal anti-inflammatory drug

4.3.2.2 Anti-inflammatory

The body immune system systematically detects a graft as a foreign body and causes inflammation. The anti-inflammatory agent, in two kinds of steroids and non-steroids, can prevent the immune response. For instance, one of the steroids named glucocorticoids inhibits cytokine-related inflammation by reducing the release of interleukin (IL)-1, tumor necrosis factor (TNF)-a, granulocyte macrophage–colony-stimulating factor (GM-CSF), IL-3, IL-4, IL-5, IL-6 and IL-8 [45].

4.3.2.3 Anti-osteoporotic

One of the bone diseases that usually encounters the aged community is osteoporosis. The disease happens when the balance between the bone resorption and bone formation is struck. Treatment strategy aims either to prevent resorption or to improve formation. The former kind of drugs includes bisphosphonates, estrogen, and calcitonin, and the latter one is comprised of parathyroid hormone. Among drugs, strontium is a drug that benefits from both of the mentioned features. Some new anti-osteoporosis agents with more potency have been recently introduced, including denosumab [46].

4.3.2.4 Anticancer

Commonly, cancer cells relocate from the original tumor site, such as lungs, breasts, and prostate, to bone tissue. The treatment is to eliminate the cancerous site and fill it with cement that releases chemotherapeutic agents. The common agents include cyclophosphamide, methotrexate, and 5-fluorouracil and newer anthracyclines and hexanes such as epirubicin, doxorubicin, and paclitaxel [7].

4.3.3   Nucleic Acids

Applying inductivity by carrying drugs and proteins on a scaffold requires a super-physiological quantity of them, which may result in some side effects and cytotoxic-ity. Using nucleic acid as a therapeutic agent appears as an alternative approach [47]. In this manner, a scaffold can carry nucleic acid molecules into the cells surround-ing the defect or enrich the molecule concentration around the cells. The molecules can transfer the gene encoding the specific growth factor into a cell, or take part in a sequence that modulates the expression of the growth factor. For instance, plas-mid DNA directly expresses its carried genetic information, and siRNA and miRNA interfere expression of a gene and modulate cell activity, as shown in Fig. 4.5 [48].

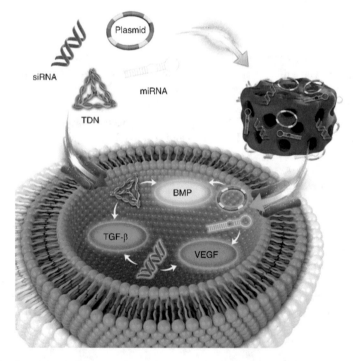

Fig. 4.5 The schematic illustrates the performance of different kinds of nucleic acid molecules, including plasmid DNA, siRNA, miRNA, and TDN, on modulating the activity of a cell (Reprinted from [48], CC BY 4.0)

Although the potential of utilizing nucleic acids is proved to be high, particularly in tissue engineering, its clinical usage is far away.

4.3.3.1 DNA

Because of the outstanding features of DNA, the therapies based on it can be known as stable, flexible, precise programmable, and simple synthesizable and modifiable. In the application of bone tissue engineering, using DNA, such as plasmid DNA, encoded with an osteogenic growth factor can supply a stable expression of the factor in the cells surrounding a defect. However, some issues are faced when transferring DNA into the cells. The issues include the large molecular weight of DNA and its negative surface charge, which may inhibit its internalization and increase the probability of its degradation. So, an appropriate carrier should be designed to deliver DNA effectively into the cells [48].

4.3.3.2 RNA

Units of genetic information, named genes, can be used either to produce a protein or to apply as various RNA molecules with transcriptional, regulatory and/or other functional activities' (Fig. 4.6) [47]. RNA possesses two different properties; it can inhibit or induce protein expression. The inhabitation happens with three types of small RNA molecules, microRNA (miRNA), small interfering RNA (siRNA) and short, and synthetic hairpin RNA (shRNA). The induction is the task of messenger RNAs (mRNAs). Employing RNA instead of DNA benefits from various aspects: easily internalization, higher transfection potential (because it is not required translocation of RNA into the nucleus), without the risk of mutation and genotoxicity, and less expensive [49].

4.3.4 Other Bioactive Agents

Peptides derived from the extracellular matrix: A substitution for growth factor is peptide that can target cellular receptors. Among peptides, B2A is known as an agent to augment spinal fusion, and P-15, a collagen-based peptide, improves osteoblast activity and differentiation of mesenchymal stem cells. In addition, a peptide, which targets BMP receptors, is studied but has not yet reached clinical usage. The issues originate from the complexity of BMP receptors and its signaling process [50].

Parathyroid hormone (PTH): PTH is commonly known for its task to control the calcium level of serum and regulating calcium-phosphate metabolism. PTH can induce osteogenesis and also its application to osteoporosis treatment is approved by the FDA since 2002. Its osteogenesis potential originates from its improvement

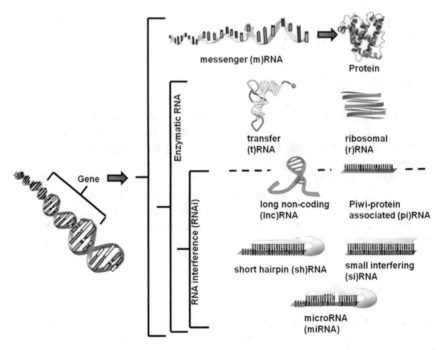

Fig. 4.6 A Gene is a short sequence of DNA molecule that is transcribed into two major subtypes of RNA molecules: messenger (m)RNA or enzymatic RNAs. The task of mRNA is the production of proteins in a process named translation. The process is subjected to the cooperation of tRNA and rRNA as well to the control of RNA interference (RNAi) molecules: short hairpin (sh)RNAs, small interference (si)RNA)s and microRNAs (miRNAs), long non-coding (lnc)RNAs and Piwi-protein (Reprinted from [47] by permission from John Wiley and Sons)

in the Wnt-beta catenin pathway, the production of proangiogenesis growth factors like VEGF and angiopoetin1 [50, 51].

Platelet lysate (PL): Two kinds of plasma-rich hemoderivatives, called platelet-rich plasma (PRP) and platelet-rich fibrin (PRF), is considered as an inexpensive source of growth factors. Although adding the agents to calcium phosphate cement improved osteogenesis potential of the cement, lack of standard preparation and test method resulted in contrasting outcome [52].

4.4   Bioactive Agent Incorporation Strategies

To improve the activity of bioactive agents and increase their half-lives, embedding them into various delivery systems is essential. The embedding process depends on the composition, architecture, and surface properties of the carriers [4]. Various methods have been innovated to incorporate different biological agents in the carrier by considering non-covalent or covalent immobilization of the agents on the carriers [53].

4.4.1 Non-covalent Bioagent Immobilization

In general, three kinds of non-covalent immobilization are introduced, including physical entrapment, physical adsorption, and ionic complexation. In the first method, bioactive agents are entrapped into the structure of the carrier during synthesis (Fig. 4.7a, b, and c), such as entrapping into a bone cement or under the coating of a carrier. In this manner, the encapsulated agents release by osmotic pressure or as a result of the degradation of the carrier. The shortage of this method is the inability to control the release rate. The second strategy, physical adsorption or physisorption (Fig. 4.7d and e), is mimicked from nature binding between a passive protein and the extracellular matrix. Ion complexation sources from binding an agent to charged carrier surface (Fig. 4.7f). These charged carriers can be composed of charged macro-molecules, such as alginate, chitosan, gelatin, hyaluronan, and also synthetic poly-electrolytes. The challenge of the last two methods is the probability of inactivation of the agents in the consequence of binding [54].

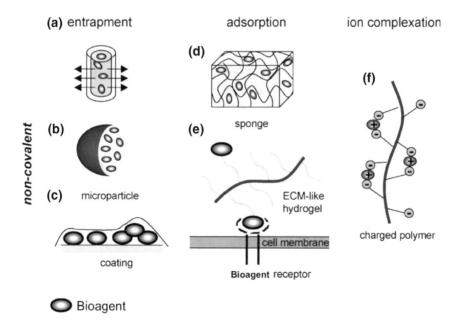

Fig. 4.7 Various non-covalent immobilization of bioactive agents in different carriers. **a–c** Physical entrapment into the structure or under the coating of a carrier. **d** and **e** Physical adsorption of passive bioactive agents to a carrier. **f** Ionic complexation of a charged agent to a charged carrier (Adapted from [54] by permission from Elsevier)

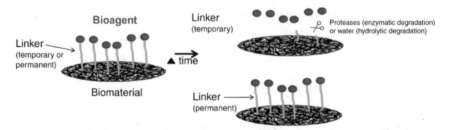

Fig. 4.8 Covalently immobilized bioagents to the surface of a carrier. The linkage can be permanent for continuous interaction with cells or be temporary to release agents via enzymatic or hydrolytic degradation (Reprinted from [35] by permission from Elsevier)

4.4.2 Covalent Bioagent Immobilization

Covalent immobilization (Fig. 4.8) is a technique that provides more control over agent release and lengthens the immobilization time. However, the process suffers a shortage that may result in the inactivation of biological agents. Nevertheless, by appropriate designing, its distribution in the structure of a carrier is more controlled and even can create a bioagent gradient [54].

4.5 Challenges in Bone Tissue Engineering

As mentioned in Chap. 1, bone tissue engineering constructs from three main elements, including scaffolds, cells, and biological agents. The strategy faces with different issues such as determining appropriate composition for a scaffold and its production method, choosing suitable kinds of the cells to achieve to a predetermined purpose, and employing single- or multi-biological component to obtain osteoinductivity. Beyond these issues (as shown in Fig. 4.9), other essential steps should be considered: preclinical in vitro and in vivo investigation, clinical trial and approval, patient expectation, and commercialization. These steps are controlled by biomaterials scientists to produce appropriate scaffold, interdisciplinary researchers to assess its preclinical criteria, surgeons who deal with the clinical part, companies which produce the product in a large-scale, and authorizing agencies to approve its application nationally and internationally [55].

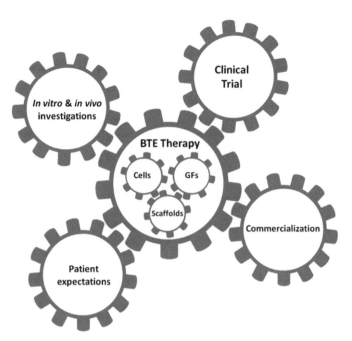

Fig. 4.9 The challenges encounter designing a therapeutic system for employing in bone tissue engineering. The challenges appear in the initial step to producing a scaffold that carries cells and biological agents and in the last step to preclinical and clinical evaluation, producing in a large scale, and obtaining appropriate approval from authorized agencies (Reprinted from [55] by permission from Elsevier)

References

1. T. Albrektsson, C. Johansson, Osteoinduction, osteoconduction and osseointegration. Eur. Spine J. **10**, S96–S101 (2001). https://doi.org/10.1007/s005860100282
2. C. Laurencin, Y. Khan, S.F. El-Amin, Bone graft substitutes. Expert Rev. Med. Devices **3**, 49–57 (2006). https://doi.org/10.1586/17434440.3.1.49
3. R.J. Miron, Y.F. Zhang, Osteoinduction: a review of old concepts with new standards. J. Dent. Res. **91**, 736–744 (2012). https://doi.org/10.1177/0022034511435260
4. R. Chen, J. Wang, C. Liu, Biomaterials act as enhancers of growth factors in bone regeneration. Adv. Funct. Mater. **26**, 8810–8823 (2016). https://doi.org/10.1002/adfm.201603197
5. T.N. Vo, F.K. Kasper, A.G. Mikos, Strategies for controlled delivery of growth factors and cells for bone regeneration. Adv. Drug Deliv. Rev. **64**, 1292–1309 (2012). https://doi.org/10.1016/j.addr.2012.01.016
6. P.S. Lienemann, M.P. Lutolf, M. Ehrbar, Biomimetic hydrogels for controlled biomolecule delivery to augment bone regeneration. Adv. Drug Deliv. Rev. **64**, 1078–1089 (2012). https://doi.org/10.1016/j.addr.2012.03.010
7. S. Qu, J. Weng, K. Duan, Y. Liu, Drug-loading calcium phosphate cements for medical applications. Dev. Appl. Calcium Phosphate Bone Cem., 299–332 (2018). https://doi.org/10.1007/978-981-10-5975-9_7
8. S. Bose, S. Tarafder, Calcium phosphate ceramic systems in growth factor and drug delivery for bone tissue engineering: a review. Acta Biomater. **8**, 1401–1421 (2012). https://doi.org/10.1016/j.actbio.2011.11.017

9. E.M.M. Van Lieshout, V. Alt, Bone graft substitutes and bone morphogenetic proteins for osteoporotic fractures: what is the evidence? Injury **47**, S43–S46 (2016). https://doi.org/10.1016/S0020-1383(16)30011-0

10. D.H.R. Kempen, L.B. Creemers, J. Alblas, L. Lu, A.J. Verbout, M.J. Yaszemski, W.J.A. Dhert, Growth factor interactions in bone regeneration. Tissue Eng. Part B Rev. **16**, 551–566 (2010). https://doi.org/10.1089/ten.teb.2010.0176

11. T. Kang, X. Hua, P. Liang, M. Rao, Q. Wang, C. Quan, C. Zhang, Q. Jiang, Synergistic reinforcement of polydopamine-coated hydroxyapatite and BMP2 biomimetic peptide on the bioactivity of PMMA-based cement. Compos. Sci. Technol. **123**, 232–240 (2016). https://doi.org/10.1016/j.compscitech.2016.01.002

12. V. Devescovi, E. Leonardi, G. Ciapetti, E. Cenni, Growth factors in bone repair. Chir. Organi. Mov. **92**, 161–168 (2008). https://doi.org/10.1007/s12306-008-0064-1

13. J. Zeng, J. Lin, G. Yao, K. Kong, X. Wang, Effect of modified compound calcium phosphate cement on the differentiation and osteogenesis of bone mesenchymal stem cells. J. Orthop. Surg. Res. **12**, 102 (2017). https://doi.org/10.1186/s13018-017-0598-8

14. F. Gunnella, E. Kunisch, M. Bungartz, S. Maenz, L. Horbert, L. Xin, J. Mika, J. Borowski, S. Bischoff, H. Schubert, P. Hortschansky, A. Sachse, B. Illerhaus, J. Günster, J. Bossert, K.D. Jandt, F. Plöger, R.W. Kinne, O. Brinkmann, Low-dose BMP-2 is sufficient to enhance the bone formation induced by an injectable, PLGA fiber-reinforced, brushite-forming cement in a sheep defect model of lumbar osteopenia. Spine J. **17**, 1699–1711 (2017). https://doi.org/10.1016/j.spinee.2017.06.005

15. A.K. Teotia, A. Gupta, D.B. Raina, L. Lidgren, A. Kumar, Gelatin-modified bone substitute with bioactive molecules enhance cellular interactions and bone regeneration. ACS Appl. Mater. Interfaces. **8**, 10775–10787 (2016). https://doi.org/10.1021/acsami.6b02145

16. G.H. Lee, P. Makkar, K. Paul, B. Lee, Incorporation of BMP-2 loaded collagen conjugated BCP granules in calcium phosphate cement based injectable bone substitutes for improved bone regeneration. Mater. Sci. Eng., C **77**, 713–724 (2017). https://doi.org/10.1016/j.msec.2017.03.296

17. S. Ding, J. Zhang, Y. Tian, B. Huang, Y. Yuan, C. Liu, Magnesium modification up-regulates the bioactivity of bone morphogenetic protein-2 upon calcium phosphate cement via enhanced BMP receptor recognition and Smad signaling pathway. Colloids Surf. B Biointerfaces **145**, 140–151 (2016). https://doi.org/10.1016/j.colsurfb.2016.04.045

18. B. Huang, Y. Tian, W. Zhang, Y. Ma, Y. Yuan, C. Liu, Strontium doping promotes bioactivity of rhBMP-2 upon calcium phosphate cement via elevated recognition and expression of BMPR-IA. Colloids Surf. B Biointerfaces **159**, 684–695 (2017). https://doi.org/10.1016/j.colsurfb.2017.06.041

19. B. Huang, Z. Wu, S. Ding, Y. Yuan, C. Liu, Localization and promotion of recombinant human bone morphogenetic protein-2 bioactivity on extracellular matrix mimetic chondroitin sulfate-functionalized calcium phosphate cement scaffolds. Acta Biomater. **71**, 184–199 (2018). https://doi.org/10.1016/j.actbio.2018.01.004

20. R. Liu, X. Wu, J. Li, X. Liu, Z. Huang, Y. Yuan, X. Gao, B. Lin, B. Yu, Y. Chen, The promotion of bone tissue regeneration by BMP2-derived peptide P24-loaded calcium phosphate cement microspheres. Ceram. Int. **42**, 3177–3189 (2016). https://doi.org/10.1016/j.ceramint.2015.10.108

21. J. Baek, H.-D. Jung, T.-S. Jang, S.W. Kim, M.-H. Kang, H.-E. Kim, Y.-H. Koh, Synthesis and evaluation of bone morphogenetic protein (BMP)-loaded hydroxyapatite microspheres for enhanced bone regeneration. Ceram. Int. **42**, 7748–7756 (2016). https://doi.org/10.1016/j.ceramint.2016.01.189

22. J.C. da Silva de Oliveira, E.R. Luvizuto, C.K. Sonoda, R. Okamoto, I.R. Garcia-Junior, Immunohistochemistry evaluation of BMP-2 with β-tricalcium phosphate matrix, polylactic and polyglycolic acid gel, and calcium phosphate cement in rats. Oral Maxillofac. Surg. **21**, 247–258 (2017). https://doi.org/10.1007/s10006-017-0624-3

23. R.O. Huse, P. Quinten Ruhe, J.G.C. Wolke, J.A. Jansen, The use of porous calcium phosphate scaffolds with transforming growth factor beta 1 as an onlay bone graft substitute. An experimental study in rats. Clin. Oral Implants Res. **15**, 741–749 (2004). https://doi.org/10.1111/j.1600-0501.2004.01068.x

24. A. Plachokova, D. Link, J. van den Dolder, J. van den Beucken, J. Jansen, Bone regenerative properties of injectable PGLA–CaP composite with TGF-β1 in a rat augmentation model. J. Tissue Eng. Regen. Med. **1**, 457–464 (2007). https://doi.org/10.1002/term.59

25. D.P. Link, J. van den Dolder, J.J. van den Beucken, J.G. Wolke, A.G. Mikos, J.A. Jansen, Bone response and mechanical strength of rabbit femoral defects filled with injectable CaP cements containing TGF-β1 loaded gelatin microparticles. Biomaterials **29**, 675–682 (2008). https://doi.org/10.1016/j.biomaterials.2007.10.029

26. S.-H. Lee, H. Shin, Matrices and scaffolds for delivery of bioactive molecules in bone and cartilage tissue engineering. Adv. Drug Deliv. Rev. **59**, 339–359 (2007). https://doi.org/10.1016/j.addr.2007.03.016

27. D.A. Oortgiesen, X.F. Walboomers, A.L. Bronckers, G.J. Meijer, J.A. Jansen, Periodontal regeneration using an injectable bone cement combined with BMP-2 or FGF-2. J. Tissue Eng. Regen. Med. **8**, 202–209 (2014). https://doi.org/10.1002/term.1514

28. R. Tsuboi, J.-I. Sasaki, H. Kitagawa, I. Yoshimoto, F. Takeshige, S. Imazato, Development of a novel dental resin cement incorporating FGF-2-loaded polymer particles with the ability to promote tissue regeneration. Dent. Mater. **34**, 641–648 (2018). https://doi.org/10.1016/j.dental.2018.01.007

29. S. Vahabzadeh, A. Bandyopadhyay, S. Bose, R. Mandal, S.K. Nandi, IGF-loaded silicon and zinc doped brushite cement: physico-mechanical characterization and in vivo osteogenesis evaluation. Integr. Biol. **7**, 1561–1573 (2015). https://doi.org/10.1039/c5ib00114e

30. A. Lode, C. Wolf-Brandstetter, A. Reinstorf, A. Bernhardt, U. König, W. Pompe, M. Gelinsky, Calcium phosphate bone cements, functionalized with VEGF: release kinetics and biological activity. J. Biomed. Mater. Res., Part A **81A**, 474–483 (2007). https://doi.org/10.1002/jbm.a.31024

31. A. Lode, A. Reinstorf, A. Bernhardt, C. Wolf-Brandstetter, U. König, M. Gelinsky, Heparin modification of calcium phosphate bone cements for VEGF functionalization. J. Biomed. Mater. Res., Part A **86A**, 749–759 (2008). https://doi.org/10.1002/jbm.a.31581

32. A.R. Akkineni, Y. Luo, M. Schumacher, B. Nies, A. Lode, M. Gelinsky, 3D plotting of growth factor loaded calcium phosphate cement scaffolds. Acta Biomater. **27**, 264–274 (2015). https://doi.org/10.1016/j.actbio.2015.08.036

33. T. Ahlfeld, A.R. Akkineni, Y. Förster, T. Köhler, S. Knaack, M. Gelinsky, A. Lode, Design and fabrication of complex scaffolds for bone defect healing: combined 3D plotting of a calcium phosphate cement and a growth factor-loaded hydrogel. Ann. Biomed. Eng. **45**, 224–236 (2017). https://doi.org/10.1007/s10439-016-1685-4

34. T. Ahlfeld, F.P. Schuster, Y. Förster, M. Quade, A.R. Akkineni, C. Rentsch, S. Rammelt, M. Gelinsky, A. Lode, 3D plotted biphasic bone scaffolds for growth factor delivery: biological characterization in vitro and in vivo. Adv. Healthc. Mater. **8**, 1801512 (2019). https://doi.org/10.1002/adhm.201801512

35. M. Mehta, K. Schmidt-Bleek, G.N. Duda, D.J. Mooney, Biomaterial delivery of morphogens to mimic the natural healing cascade in bone. Adv. Drug Deliv. Rev. **64**, 1257–1276 (2012). https://doi.org/10.1016/j.addr.2012.05.006

36. J.A. Jansen, J.W.M. Vehof, P.Q. Ruhé, H. Kroeze-Deutman, Y. Kuboki, H. Takita, E.L. Hedberg, A.G. Mikos, Growth factor-loaded scaffolds for bone engineering. J. Control. Release **101**, 127–136 (2005). https://doi.org/10.1016/j.jconrel.2004.07.005

37. E.A. Bayer, J. Jordan, A. Roy, R. Gottardi, M.V. Fedorchak, P.N. Kumta, S.R. Little, Programmed platelet-derived growth factor-BB and bone morphogenetic protein-2 delivery from a hybrid calcium phosphate/alginate scaffold. Tissue Eng. Part A **23**, 1382–1393 (2017). https://doi.org/10.1089/ten.tea.2017.0027

38. R. Reyes, B. De la Riva, A. Delgado, A. Hernández, E. Sánchez, C. Évora, Effect of triple growth factor controlled delivery by a brushite–PLGA system on a bone defect. Injury **43**, 334–342 (2012). https://doi.org/10.1016/j.injury.2011.10.008

39. R.P.F. Lanao, J.W.M. Hoekstra, J.G.C. Wolke, S.C.G. Leeuwenburgh, A.S. Plachokova, O.C. Boerman, J.J.J.P. van den Beucken, J.A. Jansen, Bone regenerative properties of injectable calcium phosphate/PLGA cement in an alveolar bone defect. Key Eng. Mater. **529–530**, 300–303 (2012). https://doi.org/10.4028/www.scientific.net/KEM.529-530.300

40. R.P. Félix Lanao, J.W.M. Hoekstra, J.G.C. Wolke, S.C.G. Leeuwenburgh, A.S. Plachokova, O.C. Boerman, J.J.J.P. van den Beucken, J.A. Jansen, Porous calcium phosphate cement for alveolar bone regeneration. J. Tissue Eng. Regen. Med. **8**, 473–482 (2014). https://doi.org/10.1002/term.1546

41. Y.-C. Chiang, H.-H. Chang, C.-C. Wong, Y.-P. Wang, Y.-L. Wang, W.-H. Huang, C.-P. Lin, Nanocrystalline calcium sulfate/hydroxyapatite biphasic compound as a TGF-β1/VEGF reservoir for vital pulp therapy. Dent. Mater. **32**, 1197–1208 (2016). https://doi.org/10.1016/j.dental.2016.06.013

42. K. Lee, M.D. Weir, E. Lippens, M. Mehta, P. Wang, G.N. Duda, W.S. Kim, D.J. Mooney, H.H.K. Xu, Bone regeneration via novel macroporous CPC scaffolds in critical-sized cranial defects in rats. Dent. Mater. **30**, e199–e207 (2014). https://doi.org/10.1016/j.dental.2014.03.008

43. M.-P. Ginebra, C. Canal, M. Espanol, D. Pastorino, E.B. Montufar, Calcium phosphate cements as drug delivery materials. Adv. Drug Deliv. Rev. **64**, 1090–1110 (2012). https://doi.org/10.1016/j.addr.2012.01.008

44. M.-P. Ginebra, T. Traykova, J.A. Planell, Calcium phosphate cements: competitive drug carriers for the musculoskeletal system? Biomaterials **27**, 2171–2177 (2006). https://doi.org/10.1016/j.biomaterials.2005.11.023

45. V. Mouriño, A.R. Boccaccini, Bone tissue engineering therapeutics: controlled drug delivery in three-dimensional scaffolds. J. R. Soc. Interface **7**, 209–227 (2010). https://doi.org/10.1098/rsif.2009.0379

46. A. Shuid, N. Ibrahim, M. Amin, I. Mohamed, Drug delivery systems for prevention and treatment of osteoporotic fracture. Curr. Drug Targets **14**, 1558–1564 (2013). https://doi.org/10.2174/13894501146661311081539 05

47. R.M. Raftery, D.P. Walsh, I.M. Castaño, A. Heise, G.P. Duffy, S.-A. Cryan, F.J. O'Brien, Delivering nucleic-acid based nanomedicines on biomaterial scaffolds for orthopedic tissue repair: challenges, progress and future perspectives. Adv. Mater. **28**, 5447–5469 (2016). https://doi.org/10.1002/adma.201505088

48. Y. Zhang, W. Ma, Y. Zhan, C. Mao, X. Shao, X. Xie, X. Wei, Y. Lin, Nucleic acids and analogs for bone regeneration. Bone Res. **6**, 37 (2018). https://doi.org/10.1038/s41413-018-0042-7

49. E.R. Balmayor, C.H. Evans, RNA therapeutics for tissue engineering. Tissue Eng. Part A **25**, 9–11 (2019). https://doi.org/10.1089/ten.tea.2018.0315

50. A. Ho-Shui-Ling, J. Bolander, L.E. Rustom, A.W. Johnson, F.P. Luyten, C. Picart, Bone regeneration strategies: engineered scaffolds, bioactive molecules and stem cells current stage and future perspectives. Biomaterials **180**, 143–162 (2018). https://doi.org/10.1016/j.biomaterials.2018.07.017

51. V. Martin, A. Bettencourt, Bone regeneration: biomaterials as local delivery systems with improved osteoinductive properties. Mater. Sci. Eng., C **82**, 363–371 (2018). https://doi.org/10.1016/j.msec.2017.04.038

52. P.S. Babo, V.E. Santo, M.E. Gomes, R.L. Reis, Development of an injectable calcium phosphate/hyaluronic acid microparticles system for platelet lysate sustained delivery aiming bone regeneration. Macromol. Biosci. **16**, 1662–1677 (2016). https://doi.org/10.1002/mabi.201600141

53. E. Nyberg, C. Holmes, T. Witham, W.L. Grayson, Growth factor-eluting technologies for bone tissue engineering. Drug Deliv. Transl. Res. **6**, 184–194 (2016). https://doi.org/10.1007/s13346-015-0233-3

54. V. Luginbuehl, L. Meinel, H.P. Merkle, B. Gander, Localized delivery of growth factors for bone repair. Eur. J. Pharm. Biopharm. **58**, 197–208 (2004). https://doi.org/10.1016/j.ejpb.2004.03.004

55. L. Roseti, V. Parisi, M. Petretta, C. Cavallo, G. Desando, I. Bartolotti, B. Grigolo, Scaffolds for bone tissue engineering: state of the art and new perspectives. Mater. Sci. Eng., C **78**, 1246–1262 (2017). https://doi.org/10.1016/j.msec.2017.05.017

Printed in the United States
By Bookmasters